Ivan Ant

METAPHYSICS
OF
ASTROLOGY

Why Astrology Works

SAMKHYA PUBLISHING LTD
London, 2020

Translated by
Milica Breber

Proofreading & editing by
James Joshua Pennington, PhD

Copyright © 2020 by SAMKHYA PUBLISHING LTD

ISBN: 9781656673336

TABLE OF CONTENTS:

Introduction 5

Some Basic Facts and Misconceptions 7

The Anthropic Principle and Human
Physical Embodiment 28

Zodiac 43

Planets or the Psychodynamics of Space-Time 56

Destiny Resides in the Matter 73

Freedom Resides in the Soul 97

INTRODUCTION

Unlike the majority of other books on astrology, we will not deal here with the teaching of the basic principles necessary for horoscope reading. Such books, both good and bad, are copious. Instead, we will deal with **why astrology works** in the first place and how its original principles came about as well as how the space–time of cosmos shapes within us into (un)known existence, into planets, into life itself, and into destiny.

To approximate this vision that encompasses the micro and the macrocosm, at least to the level we are capable, we will utilize some of the achievements of theoretical physics. Astrology has been known as the queen of all sciences from ancient times. Its zodiac circle is like a whirlpool where all existential experiences are acquired. We will not mention them here, but instead focus on the most recent ones, with the added contribution of the experience of personal insight. We will stick to the fundamental theories of physical reality here, in an attempt to clarify the essence of astrological influences. There are the theories of the holographic universe and the Strong Anthropic Principle, which are complementary and claim that the universe, in its essence, exists to shape the conscious subject – in other words, us. It is my understanding that astrology shows every detail of this shaping for every person individually. Therefore, the emphasis on personal experience should not act as a deterrent but should rather do its best to stimulate, given that this is the only correct perspective that can make any sense out of astrology. If it shows in detail how from the wholeness of nature a personality is formed, its purposefulness cannot be expressed otherwise, but as personal experience only.

The fundamental principles of astrology put forth in this book follow the ancient teaching of Samkhya. This teaching reveals the true relationship of the human essence, or soul (Purusha), and nature (Prakrti). The soul is a transcendental witness, and nature is what acts as the emergent universe. Its effect is such that it seduces the consciousness of the soul, making it identify with its phenomena and functioning. Astrology shows us all every detail of this seduction and the conditioning the soul experiences for each person separately. Prakrti is best experienced via astrology. This brief discourse (written about 20 years ago) should, therefore, assist us in our task of differentiating between our essence or soul and all possible natural influences.

IVAN ANTIC
AUTHOR

SOME BASIC FACTS AND MISCONCEPTIONS
♄ ☋   ☊ ♆

At the moment a newborn baby draws its first breath, the spirit of this world enters it and gives it life. This is the spirit of time; the moment the baby inhales. Because of the unity of nature, every moment is the embodiment of all existence. It is shaped by the whole: the cosmos, the position of the earth in relation to the sun, neighboring planets, and faraway stars. All those bodies are energy forms. All of nature is energy, which expresses itself by shaping itself in various forms, and at the moment when an infant cuts its physical ties to its mother, its connection to the new, wider whole is introduced – its connection to the cosmos. With its initial breath, it takes the energy of the entire cosmos into its lungs. While in its mother's womb it simply existed, as its body was being formed. In its newborn environment, it will have to actively participate in its existence, as the entirety surrounding it will form its personality. After its birth, a child enters the drama of all possible life experiences, full of events. In their essence, these events are motion and movement reflecting the true nature of the cosmos, which is represented by general and perpetual movement.

Astrology shows us how this movement takes place following the law of the golden ratio, the sacred geometry.

With the first breath, the human receives the pattern of their personification, their character, which, over time, becomes expressed as destiny reflected in detail by their natal chart. In the mother's womb, the human body is formed. Once it emerges into this world, the psyche will form through action. The same way the mother's body formed the infant, the movement of the meaning of events of this world will form the infant's psyche. Astrology shows us how this world has been constructed and what generates its events. These events are the destiny of all beings. The destiny of every human is to become an independent person by becoming aware of the whole that sets everything in motion and attains a state of utter authenticity, the state of Self or the consciousness of their soul, which means conscious unity with the whole, including all misunderstandings and flaws in horoscope interpretation when something else is expected.

A completed personality is a personification of the whole, which creates everything at every single moment, and therefore humans cannot be taken to their completeness any other way but by the whole itself that made them. The road to such completeness and self-knowing as the knowledge of the whole is always individual because everything in nature is unique. For this reason, the road to Selfhood is a process of individuation.[1] It is unique for everyone, and its uniqueness is determined by the time and place of birth, by the intersection of space and time. The birth of a human, whose mission is to grow into a personality, is nothing but the intersection of space and time. Once they intersect, a conscious life is born. In all other areas of existence, this intersection is not the real

---

[1] On the process of individuation see Carl G. Jung: *Man and His Symbols*.

thing; they are all twisted, incomplete forms that constitute all other beings and everything living in nature other than the human. Only in the human image does the proper intersection of the time horizontal and vertical of eternal space, the heaven and the earth, the consciousness and existence, take place. The human is the only complete form of existence; the spiritual embodiment of the whole because the spirit of the whole, which creates everything, can attain the direct self-knowing through it. In all other forms, it simply happens as the manifestation of nature itself.

Just as space–time has its pyramidal hierarchy (at the top, it narrows to consciousness and freedom, and toward the base, it widens to ever-growing inertia and conditionality), the formation of the personality, too, has its pyramidal structure. The widest base of the pyramidal structure of the personality is determined by the year of birth. Each year is determined by one of the cycles of the activity of the sun. There are twelve of them, and their phases matches the lunar cycles. Character traits of all individuals born in the same year are depicted by the Chinese Zodiac in detail. The year of birth is the most general reference of an individual character that has been determined following the cycles of solar activity. ***An astrologer who fails to take Chinese Zodiac into account while analyzing a certain chart is unable to view its completeness.*** They dabble with details only. To recognize this widest foundation of a certain personality, a more detailed and time-consuming observation is required; it is not obvious at first glance. The old Western astrologers, who were unfamiliar with the Chinese Zodiac, suspected that there must be yearly cycles and tried to establish them by adding the significance of a certain planet to every year.

It is a well-known fact that lunar astrology relates to the Eastern peoples, whereas solar astrology relates to Western ones. It is in accordance with the different mentality of the people of the East and the West. The Chinese Zodiac is lunar, but the yearly cycles it represents may refer to the Westerners as well, or more accurately, it affects all people in the world, not just the Chinese. This is further proof that the characteristics of yearly cycles have a global impact and are not connected with any specific astrological tradition. Lunar cycles merely coincide with them.

The following, higher structure of the differentiation of personality is determined by the season or the month of birth. The Western astrological tradition shows this in detail. The issue here is that it describes the position of the earth in relation to the sun and the other planets.

When observed from the earth, the sun moves along its path, ecliptically and unevenly, throughout the year. Due to the earth's rotational axis wobble, it is found in its lowest point in the horizon in the Northern Hemisphere at the beginning of winter, on December 22, which marks the shortest day, and during the summer, on June 22, in its highest position, which marks the longest day of the year. It then begins to wane again, and at the moment it intersects the middle of its extreme positions, September 22 and March 22, which is the line of the equator, autumn and spring begin, which are equinoxes. Therefore, the beginning of each season is determined by the intersection of the ecliptic and celestial equators (a projection from the earth).

This was taken to be the beginning of the cardinal Zodiac signs: Capricorn, Cancer, Libra, and Aries. Those are the four cardinal points of the Zodiac, and they have always been the basis for measuring the time all over the world. Starting from them, all other signs were determined, set 30 degrees apart from each other.

What is at work here is *time*, within the boundaries of which all phenomena of the wholeness is shaped; or more accurately, the ***position of the earth in the space of cosmos and not the mythology associated with certain constellations***. The names of the signs have been adopted from the constellations, but the signs themselves do not correspond to the constellations due to a slight oscillation of the earth's polar axis, the point of intersection of the ecliptic (sun's orbit) and the equator. Therefore, the moment when the sun crosses from the Southern to the Northern Hemisphere, when spring begins, does not happen when the sun enters the constellation by the name of Aries. This point of intersection of the sun's orbit across the celestial equator continues shifting slightly retrograde, approximately one degree every 72 years, and this is called "precession." Nowadays, it is at the end of the horoscope sign of Aquarius, which means that precession passes through one sign for 2,160 years, or for the duration of one astrological era. The signs and the homonymous constellations corresponded around 300 BC, but they were never before or after that date congruent.

To those whose knowledge of astrology is very limited, this non-alignment offers crucial proof for overthrowing the credibility of astrological theory and practice. Anyone who is minimally familiar with astrological principles is aware of the fact that ***signs are not constellations***; only the names are the same, and the signs are constituted by certain ***positions of the earth in relation to the sun*** as well as the concrete, real changes in nature generated by those positions: the four seasons.

Therefore, we are not affected by some mysterious rays from the faraway constellations, filled with the fancy of our ancestors, but a real position of the earth regarding its environment and the state of nature at that moment. It is what keeps conditioning our lives in a way that can be validated both meticulously and scientifically, something that astrology has been doing since time immemorial. The reason why it still has not been widely accepted lies in the fact that neither the human personality nor the meaning of the Self has been able to gain public acceptance and recognition to this day. Only recently has there been sporadically recognized the necessity for appreciating the elementary rights to the biological survival that are very boastfully proclaimed as democracy, human rights, and liberties. However, they have not yet reached the peak of civilizational achievement, although they represent nothing but the basic prerequisite for a normal biological life. We still have a long way to go before accepting and fully appreciating the sense of wholeness of the human being and personality astrology speaks of. It points to the unity between the human and the cosmos, whereas the human today is an ununified whole, especially when it comes to immediate family.

With its seasons, nature shapes human character. Even for a layperson, it is pretty evident that someone born in winter (Capricorn) with their reserved, somewhat cold and abstract character, is different from another born in summer or at the end of spring (Gemini, Cancer), who is emotional, warm, communicative, and accessible; those born in the beginning of spring (Aries), with their brisk energetic approach to life, are considerably different from the hesitant, moderate, and considerate character types born at the beginning of fall (Libra).

The next, finer differentiation refers to the decans. Each sign has three decans of 10 degrees and one planet to rule symbolically over each one. This means that each sign has three phases of its influence: the initial, middle, and final one. Those born under one sign are, according to this, different by the decans their birth falls under.

All decans have been assigned the following planetary influences:

Aries: Mars, sun, Venus
Taurus: Mercury, moon, Saturn
Gemini: Jupiter, Mars, sun
Cancer: Venus, Mercury, moon
Leo: Saturn, Jupiter, Mars
Virgo: sun, Venus, Mercury
Libra: moon, Saturn, Jupiter
Scorpio: Mars, sun, Venus
Sagittarius: Mercury, moon, Saturn
Capricorn: Jupiter, Mars, sun
Aquarius: Venus, Mercury, moon
Pisces: Saturn, Jupiter, Mars

There is an even finer differentiation according to the signs of the Zodiac. Namely, it is assumed that each degree of each sign has a special impact. It is pointless to debate on this subject here, for it would take up too much space. The pyramidal structure of space–time, with which the whole human personality is formed, we will finish on the finest level, and that is the moment of birth, the hour and the very minute of it. It is determined by the position of the earth's rotation in relation to the sun, and it is the time of day or the exact position of the sun concerning the place of birth.

To a careful observer, the one who has not become acquainted with astrology, it may be apparent that the character of a person born at noon (sun in the tenth house) is different from the one born at midnight (sun in the fourth house). The former is turned toward the public domain, career, politics, and the events of the world, and the destiny unmistakably leads them there, while the midnight type is attached to their family and the home front, prefers to sit at home rather than travel all the time, or is in some way forced to be at home or do the homemaking. Their character is more intimate than the former.

It is even easier to spot the difference between someone who was born early in the morning, just before sunrise (the first house), from the one born when the sun sets (the seventh house). The former one is very self-conscious and self-sufficient, while the latter one is dependent on others and the environment; the former one enjoys their solitude, and the latter one seeks the company of people, continuously wanting to present themselves in public. The former is more inclined to experience, whereas the latter one is likely to express. Similarly, nature determines the character of an individual using part of day or night.

The moment of birth, as the peak of the pyramidal structure of personality, provides the closing and defining imprint of the entire structure. Ascendant is determined with it, as well as the position of all twelve houses that determine our destiny, which is once again, nothing but the way in which an individual will be formed.

One of the most common misconceptions of laymen about astrology is the conviction that some distant celestial bodies cannot affect the events of their lives, because they have their own "free will," at least to a substantial degree if not entirely.

The initial nebulous assumption is the notion of the distance of celestial bodies astrology speaks of, which goes to show the level of ignorance of the astronomical vastness of space we live in. If we were to compare our galaxy with the human body, the distance between Pluto and us would be no greater than the neighboring brain cells in our body. Their organic ties could not be disputed by the layman even, but the astrological assumptions of the unity of nature will soon be forgotten, regardless of the fact they are only too well aware of the burden of evidence. Astrophysicists are particularly eager to play this game. Of all average people, they know the best that gravity is the most powerful pull in the cosmos. It gives cohesion to the planets and keeps all cosmic systems together in perfect harmony, from the solar to galactic level. However, despite this fact, they will conveniently become ignorant of the fact that the same gravity, the teaching which astrology is based on, plays the key role in organic life, too. Moreover, it does not require too much knowledge to realize that all organic life originated under the influence of gravity, the sun first and foremost, but the moon and all other planets as well, because all bodies have gravity. Organic life is, in effect, our daily life and the basis for our destiny to take shape. In fact, **organic life is our destiny** - overcoming it falls under the category of the ideal of human perfection and is known as the experience of transcendence in all spiritual traditions. However, this ideal is not the object of desire of ordinary people conditioned by their destiny, which boils down to everyday work and reproduction, together with rest, such as entertainment and the accompanying trivialities. This is all intrinsic to organic life. If they happen to own up to the fact that gravity plays a part in organic life, perhaps observing that it ties them to the ground and that the rhythm of life depends on the gravity of the earth, moon, and sun (and even the gravity of some other planets they are still not ready to admit), they will still not agree that organic life is the same as destiny - because they are unable to distinguish between what destiny and real freedom are – an issue that occupies an important place in this debate. When the earth is observed from the cosmos, it looks like one whole living being, where lives, movements, and the destinies of billions of human beings on it appear like vegetative functions of its organism - the tiny bodies of astrophysicists representing bacteria, which obstruct the functioning of it with their ignorance although they live at its expense.

The further assumption in this delusion is to know what phenomena are, the events that constitute our life. This "knowledge" is typical for the unconscious human, who sees natural processes as something different from themselves and from what they do. Even the most educated among the unconscious people (physicists) are prone to it even though they know that macrocosmic movement is subjected to the strong gravitational pull and microcosmic to the strong nuclear and electromagnetic forces, everything that is between these two extremes, which is our movement in everyday life. Despite this logic and these facts, they consider themselves to be different from the remaining nature, although our world is between these macro and microcosmic forces as if between hammer and anvil.

Only at the point between the micro and macro world, in our human endeavor, is the place where our "free will" rules; although almost no one seems to brag about having used it properly. This is how the human ego maintains the illusion of its survival.

In reality, everything that exists in any way possible, in or outside us, in this or any other world, is a constituent of the conditioned natural processes, and there is no such thing as "unnatural" phenomena. It is different in the level of vibrations and the quality of invested energy. Those differences give various forms to everything that exists and happens. The entire nature is made up of one energy, and it manifests itself as a movement. The movement of nature is the same everywhere; there is no difference between the movement of electrons and planets or our movement through life, because energy is the same everywhere. However, its implementation gives a different character to the phenomena. Beyond that, there is no difference between solid matter and events, the outer or the inner ones. An illusion that there is a fundamental difference between the invisible processes of the events of destiny, the visible objects, and our being, we can only interpret as our attachment to the limited physical senses and their point of view. Forces that lead humans through life and give them a certain destiny best reflected in the events that occur are as much a part of nature as the force that keeps the blood flowing through our body and all of the organic life. Let us keep in mind that everything we are prepared to do and have already done is to put food on the table and pay our bills. The egoic way of observing things is the only reason why we consider our blood flow to be a natural urge; our external actions, our "free will," and all the other events are a mere coincidence or destiny. Such distinction is the rampart of our ego and ignorance, and thereby our suffering. However, in nature, there is no suffering. It is a divine, perfect whole. Suffering exists only within the boundaries of ego in the outside world, which the ego projects. We do not want to see the unity of nature in organic survival and our deeds, because we do not want to admit to ourselves that we have done very little in our lives outside satisfying our biological needs, barely scraping the surface of our spiritual essence – the real reason why we exist in the first place. This is largely due to the traditional ignorance of what our spiritual essence is, and we will delve into this subject further on.

The last and firmest assumption of a layperson who "does not believe in astrology" is based on the previous one. By distinguishing the natural processes from their actions, their ego creates an illusion that they have free will. This conviction, which is nothing but the manifestation of their unconsciousness, is so widespread that many of the commercial astrologers fell in line with it. They go so far as to maintain that the human has free will and astrological factors only show tendencies, which a person can either use and benefit from or pass on and end up failing in the long run. It is true, but only for an enlightened human; not for the mere mortal. This conviction is appealing because it confirms the egoic unconsciousness, although with just a little common sense, it is obvious that humans cannot breathe or blink at will and that all of their physiological functions are unconscious and spontaneous. This is not only true for the internal processes but the external events as well. It is hard to be fully aware of this because the natural conditionality is flexible enough to enable all the life and movements the way we know them within the boundaries of our ego. However, at the same time, it does not facilitate cognition of other aspects of life that are far more meaningful, bigger, and more miraculous than the things we are now familiar with. Every mature human, upon recapitulation of their past life and actions, would admit to not being the one who chose their destiny, at least most of the events of their life. The recognition of this state is further complicated by the fact that the things they consider to be the expressions of their free will are predetermined by astrological factors at birth. The planetary aspects and positions that make people dedicate their lives to sports or theology, for instance, humans accept as an expression of their own free. This is because astrological factors work from within, unconsciously, much like "our" desire and "our" will work in the outside world in the form of events. They predetermine everything: whether someone will be religious or not, superficial or fanatic, or successful in their career or a failure in life as well as the kind of marriage a person will have and what clothes he or she will choose to wear.

Everybody exercises their will, but it is different from one individual to another because everybody's will is determined by the astrological influences from their natal chart or the moment of birth. These influences are expressed by everyone as their will, but it is, in reality, their temperament and character that is being expressed. If the astrological influences (aspects and the accompanying factors) are constructive and positive, they will, following their will or innate nature, experience their positive accomplishments. If they are negative and destructive, their will can only lead them to their own demise. This is why it is common knowledge that people who are experiencing their darkest hour in life, or are otherwise prone to doing bad deeds, are hard to be influenced or able to mend their ways. They see their lifestyle as an expression of "their will" and "their achievement." This complication is very well known to psychotherapists. To such individuals, the only way out of their doom is to listen to other people's advice and not to follow their innate nature or what they perceive to be their will - if that is at all possible. Because of such individuals, social and religious authority were originally introduced (where one has to commit oneself to God's will), along with ethics and morality.

The conditionality of nature demands subordination. Some events are determined to a greater degree, whereas some are far less so. Insignificant and petty events are determined in their local environment by the immediate causes or are a question of coincidences because their importance bears very little leverage on the person's life. However, life-changing events, which are fundamental to development and destiny, are determined by the higher causes. Astrology deals with the latter type. (Truth be told, even the most insignificant event could be directed by destiny) Therefore, to understand astrology, we should distinguish between relevant and irrelevant events.

To understand the law of causality that rules over nature and our destiny, it is imperative to know that nature is not only what we can perceive with our physical senses. It is multidimensional, while we can observe only a single dimension with our senses, a very narrow area of nature that we call the physical, three-dimensional world. The law of causality, however, extends over the whole nature, across all of its dimensions, which explains why we do not see the law in its entirety but only the parts that appear in passing through the physical plane, and if we happen to be nearby, we can detect them or be informed of them in some other way. Since we fail to see the entire flow of causality but only the fragments of it, the ones that resonate with our paradigm of reality appear unconnected, the mere result of accident. It seems that evil rules and there is no justice and that strokes of luck are undeserved. Not being able to perceive (hidden away from our senses) the chain of causality on the higher dimensions, events and phenomena may seem like the numinous whims of some God or devil, but miraculous events are nothing more than the manifestation of higher dimensions in the lower ones.

The reason religions and philosophies are even built on such ignorant lies in the fact that we do not see the whole of the natural processes in all dimensions but only a fragment that is accessible to us owing to our senses. (Science deals with it and therefore astrology cannot be one of its subjects, because it points to all phenomena across all dimensions.) It is as though we attempted to observe the motion of a big wheel through moving through a narrow slit, where we could only see a segment of the total happening. We would see some indefinite pieces appearing and disappearing, some cycles of this motion at best. Only when we take a few steps back and see the big picture, we realize that unconnected and inexplicable phenomena are the building blocks of the entire whole. The same applies to nature and its phenomena. We observe our physical lives through the narrow channels of our senses and limited minds, and to us, they appear imperfect, wrong, illogical, and unhappy, like victims of all kinds of suffering and injustice. It appears that life shows up anew, only to disappear into oblivion the next moment. Only when we can objectively perceive all dimensions of nature are we able to see that our physical life is a tiny piece of a far bigger process of the maturing of the soul until it reaches eternal life, which is the reflection of the perfection of nature. With its dimensions, it represents the perfect, timeless whole and purest benevolence.

Astrology displays the processes of causality that go beyond our sensory range, enabling our full comprehension of the completeness of phenomena and making sure we avoid the illusion of suffering.

The strings nature binds us with while conditioning us are long enough, fine, and invisible to provide us with the freedom of movement (although there are people who have been denied even that) necessary for satisfying the biological urges for survival, even the lowest urges drives at that. This kind of freedom is needed for biological survival, and nature has granted it, even more so that even animals are able to have it. Those individuals whose minds are restricted to the sensory and physical existence only, this dominantly biological freedom to fight for survival, create an impression that people are completely free with it, so much so that they can build democracy, human rights, and religious morality while at the same time thinking this freedom came from God. They constantly fight the negative aspects of the freedom in question as well as those people who have an opinion contrary to their own. As a result, "democracy" and "religion" fight more dirty battles for territory and survival than animal predators would.

The choice between good and evil that nature gives us simply by being born, for the sake of biological survival, is not freedom but the choice of how to satisfy the instinct for survival, which is polarization at best of the same conditionality that is flexible enough to enable the physical subsistence of an individual. We may be in a position to freely decide how to go about our urges, but we begin to project this kind of freedom that does not surpass the conditionality of nature naively onto all of the areas of life, even onto the spiritual essence we have not even encountered yet. Just because we can move to and fro, hit or caress someone (in both cases with an intent to stay safe), and talk gibberish if we choose to, this does not mean we are free. It is an illusion; it is the freedom of the ego and unconscious beingness. True freedom has yet to be attained with spiritual maturity and conquering this type of freedom of instincts that continues to enslave us with its illusions.

Astrology shows all ways in which the human is conditioned, helping us get in touch with our real spiritual freedom, the consciousness of our transcendental soul. Conditionality may be both positive and negative because nature works in all ways, through oppositions, in a stimulating and restricting way, always tempting us and making sure the lessons have been learned well. These actions thereby affect several lifetimes due to the laws of karma and reincarnation. For this reason, sudden death in youth or a hard life must be viewed as an episode of a far wider and bigger process than a single life can be. Once this type of completeness of nature and its functioning is overlooked, together with its unity of discrepancies, the interpretations of the "good and evil" set in, where the evil mostly overpowers the good.

Only the wholeness of existence is pure good; our spiritual essence is what enables it, our Self or the consciousness of the soul that is a reflection of the divine consciousness. The fragments of existence are always opposed, and we will suffer because of them for as long as we are fragmentary (residing in ego). Suffering is the unconsciousness of the whole, and we express it most often through a reaction, through the conflict with its various pieces, the people and beings that seem as "others" to us because we are alien to ourselves too. Therefore, we are able to commit evil acts only when we are unconscious, and we maintain the status quo through our unconscious reactions, while love and goodness are always the expressions of our consciousness of the unity of being and spiritual freedom, which defines human authenticity. We cannot know the truth about the being without love toward everything that happens because such "knowledge"" is merely an outer phenomenon, much like we cannot love in a phony way with the feeling of separation or without the unity with our loved one. Good and bad are not outwardly advents. They are our property that manifests our level of consciousness and unconsciousness. It is evident that the human is capable of both. When we project them outwardly (good into God, and evil into the devil), we then merely avoid taking responsibility while implementing both, thus proving to be unconscious of our habits and ways. Once we accept responsibility, it always leads us to awareness and goodness, which is universal and does not depend on morality and laws that are determined by the local environment.

Nature exists with an intent to form the conscious subject. Apart from all other life forms where it also takes place, this process is most intensively crystalized through the human being and destiny. Humans are the final act of this process. For this reason, it is the hardest for them, but they are the closest to the outcome. Astrology, as the queen of all knowledge, shows in great detail the totality of natural occurrence and the level of individual conditionality, although not to rub the nose and prove that no human is intrinsically free from the natural causality but quite the opposite - to provide them with this knowledge to be able to use to it to set themselves free by becoming aware of the wholeness of beingness, which is their Self, fundamentally. Therefore, the human is conditioned, but not irreparably. They are capable of becoming free only by realizing what it is that keeps them conditioned. Astrology may play the key role here following the logic that a knot can be untied only by learning how it came to be tied in the first place. For a human to use this knowledge properly, they must become aware of their conditionality first and not go on deceiving themselves that their freedom and independence was granted to them by simply being born into this life or through some God. ***Freedom cannot be granted; it is a characteristic of personal maturity, whereas maturity can never be a characteristic of someone's personality if it has been granted from somewhere.*** The very act of physical birth can, by no means, stand out from the other conditioned processes of nature. It places humans on the same level together with animals and plants. Their true, spiritual birth has yet to happen when, through cultural evolution and complete awareness, they become a unique and complete person in themselves, ***independent of external influence***. The more they are integrated as an individual, the more they are independent of the gravity of natural conditioning.

The issue of determining the level of free will is generally not easy but is especially difficult in astrology. The solution is simple. ***There is destiny with predetermination as well as the presence of free will, together and side by side.*** Predetermination imparts stability and causality to events, whereas freedom produces creativity and new experiences, shaping all events. Without freedom, existence would be like a prison with no life inside. Without predetermination, freedom would slip into chaos and rampage with innumerable coincidences.

Conditionality and destiny, with its predetermination, exist to the degree we are unaware of our true nature, the consciousness of our transcendental soul that exceeds nature because it enables it. The consciousness of our souls is a reflection of divine consciousness that enables the entirety of nature. Therefore, only to the degree that we aspire to strengthen the consciousness of our soul in this world and in our body can we become liberated. Therefore, conditionality exists in this world but is not complete. ***There is only one way out of conditionality, which is its transcendence in the consciousness of the soul, in self-knowing.*** The more we strengthen this consciousness, we become true masters of our destiny through liberty. When we follow our earthly and bodily urges and desires, we always act unconsciously; we are slaves of destiny. Only when we act consciously in the best interest of our soul, our conscience, can we truly liberate ourselves. Only with actions such as those can we begin to construct our freedom and the human world. ***The only human freedom is to know one's self.*** Everything else is slavery and suffering.

The only real foundation for the human cultural evolution is provided by astrology if it is used in accordance with its purpose, which is human spiritual enlightenment, and not for the fulfillment of personal desires. This type of misuse is the only reason for the many mistakes and misunderstandings. A natal chart shows the structure of our whole being to be able to provide us with the context for going beyond, into the freedom, into the consciousness of the soul. Transcendence of the conditioned beingness is the essence of the overall culture humans have created, our spirituality and religiousness. Human conditionality is best portrayed in astrology, a special code for every single human being according to the time of his or her existence, hence in a dynamic, living fashion and not metaphysical or mythological.

Astrology clearly shows how nature shapes the conscious subject and, through a series of temptations, leads them to grow into a complete, integrated personality that

will be the embodiment of the sense of the beingness of nature. Astrology will be recognized in a human society exactly to the degree that this society can recognize and acknowledge a whole and free personality.

Astrology displays the essence of our character, which is outwardly expressed in the form of destiny. In doing so, it contributes to the true understanding of our position in the world, the true cause and meaning of everything that happens to us, so that we do not blame it all on God, or the devil for that matter, or those near and dear to us. With such knowledge, we relax and open up our hearts for all beings through the insight that we are equally conditioned by nature until we attain the true spiritual awakening the conditionality drives us to. One of the most beautiful depictions of the meaning of astrology is in Manichean sources (Hegemonius, Acta Archelai, 8. Bar Khoni, Scholia, 315, 22–27):

"When the Father of Life saw how the soul suffers in the body...he called for the Paraclete. Having come he undertakes to prepare everything he needs for the task of saving souls. He made a wheel with **twelve** dishes... Put in motion by the circulation of the celestial spheres, this wheel seizes the dying souls... And the raft continues filling the dishes with souls, taking them on and then disembarks them...into Eons, where they remain in the Pillar of Glory by the name of *Perfect man*... It is a pillar of light, for it is filled with purified souls."

THE ANTHROPIC PRINCIPLE
AND HUMAN PHYSICAL EMBODIMENT

Modern theoretical physicists have invented many concepts concerning the nature of physical reality; ultimately, these all end at the same unique paradox. On the one hand, experimental breakthrough in the field of subatomic physics has dissolved almost all of the former beliefs on material reality and their solidity and consistency and have provided us with such an image of reality where everything is just an illusion orchestrated for us by the nature of sensory perception and not an expression of the reality itself. The essence of matter or the subatomic reality have proved themselves to be highly relative and directly dependent on the subject experiencing it. Soon after its occurrence, in the 1920s, quantum physics got itself entangled in the problem of the relationship of the subject and nature because the intricacies of that relationship proved to be of key importance for achieving results of experimental study. With insights of that kind it, on the other hand, took one step closer to the ancient knowledge on the essence of reality, the mystical cognitions that are at the base of all the greatest spiritual traditions.

According to Fritjof Capra, who in his acclaimed book, *The Tao of Physics*,[2] brought forth the parallel be-

---

[2] Fritjof Capra. *The Tao of Physics: An Exploration of the Parallels Between Modern Physics and Eastern Mysticism*, 1975. Shambhala Publications.

tween the modern theories in physics and ancient, mostly Eastern, mystical insights, this merging consists of eight crucial points.

1. The unity of all things and events - the world cannot be reduced to independent, smaller units.

2. This general unity always involves the conscious subject (human, observer); they are a direct participant to the point that the knowing of unity is their destiny. Everything happens to ensure the crystallization of their consciousness.

3. In this unity, all differences and oppositions are relative; opposites are the poles of unity and enable the dynamics of life.

4. All notions are fabrications of reason and cannot be applied to the very reality of nature; they do not relate to the characteristics of the reality itself.

5. Beyond the space and time of relativistic physics, there is a higher, timeless dimension, an implicit order in which all events are interconnected but not in a causal way.

6. Cosmic unity is dynamic and inseparable from its multiple occurrences, the nature is in dynamic balance, and multitude is an illusion of the limited sensory point of view.

7. Space and particles are not strictly divided and opposed but are in a dynamic and living process of perpetual movement, induction, and destruction. Particles are generated from the field oscillations that are shapeless and indeterminate and disappear within; although they are unmanifested, the fields are a reality that produce entities we observe as particles that compose the physical world.

8. The entire reality is a continuous cosmic dance of energy, a game of appearing and disappearing. Par-

ticles that constitute the physical reality are, in their foundation, energy processes and not solid objects.

Classical physics has dealt with the mechanical description of the world from Newton to the beginning of the twentieth century. It corresponded to the simple sensory experience, and it was successful in technical application. The technological revolution and modern age began when humans became aware of all aspects of motion in the sensory, physical world.

During the 1920s, deeper and more revolutionary insights into the nature of physical reality and the subatomic world came to light. They indicated that elementary particles, or the atoms that the physical world is made of, are quite different from the classical idea of matter as their behavior is that of paradox; they skip laws of cause and effect, even the limitations of space and time. Their position and speed cannot be determined precisely, because they are high in energy and a vibratory state in the form of a tiny field called the quantum field. The quantum field fluctuation decides what kind of particle or wave will be created there. Unlike the rigid mechanistic view of the world, quantum physics has demonstrated that deeper reality is more like an organic model because the observer becomes a participant of the phenomena, which equals the mystical insights on the unity of nature astrology is based on. In short, a modern physicist has come to the point of an insurmountable obstacle, which is the self. At the heart of nature, they see a reflection of the human "I."

During the early stages of quantum physics, it was clear that the results of experimental studies depend on the interpretation of the observer, who sets the parameters of the macroworld to match their microworld, which is not feasible, hence the paradox and the idea that objects in the microworld have properties of both waves and particles. This practice was later abandoned with the exception that some of the theorists[3] attempted to design a new situation on the ethical and existential plane in order to define the new science. However, the latest discoveries took this problem to the astrological dimension. It has recently been discovered that the very presence of the observer determines the microworld's response, whether it will act as a particle or a wave. It was established that the very presence of an observer would cause the quantum field to act as a particle, while without the observer in sight, it acts as a wave. It appears that the particle, or nature, knows it should assume the three-dimensional form only in front of the observer. Therefore, it is no longer the issue of interpretation of the observer, the way it was thought until recently, but its very presence.

The implications of this discovery are far-reaching and will surpass the very discovery of the subatomic world and its characteristics because the ultimate essence of nature and human existence will merge in a way never experienced before. It has, however, always been united in mystical experiences of spiritual illumination and the ancient knowledge of astrology, but our intention here is to come closer to this from the viewpoint of modern knowledge of physical reality and humanity.

The safest step in that direction will be the cosmological theory of the strong anthropic principle (SAP). In numerous results from the spheres of cosmology and quantum physics, a whole series of matches was discovered between the numeric values of some fundamental constants of nature, such as the ratio between proton

---

[3] Werner Heisenberg, Jürgen Habermas, Stephen Toulmin, David Bohm, John V. Davis, Fritjof Capra.

mass and electron mass; if the gravitational mass of protons were any different, there would be no stars. It has been observed that the very probability of life occurrence depends on these congruences, that the fundamental characteristics of the universe must be exactly as they are in order for the evolution of life, which is based on carbon, to take place and, at long last, the human as the conscious observer. This fact predisposed the forming of the strong anthropic principle, where the visible properties of space, the way they are in everything, are not a product of coincidence or natural selection between a number of possibilities but are rather a consequence of a completely definite purpose: the creation of conditions for the occurrence of a conscious subject.

The logical answer of this kind for the congruence of the many relationships that constitute the foundation of nature was not satisfactory for the renowned astrophysicist Stephen Hawking. He dismisses it in his book, *Black Holes and Baby Universes*, because he finds the statement that things are the way they are simply because we exist inadequate. He concurs that the solar system is necessary for life, however, many parts of cosmos are not suitable for any kind of living. Additionally, he claims that we are so petty and inconspicuous as compared to the sheer size of the cosmos, which may be an oversight on his part because magnitude acts as the guarantee for the many suitable places of the kind of life that we have. Apart from the fact that he finds this theory "hard to accept," he offers no plausible evidence to suggest otherwise.

Being on the quest for the big uniting theory, the one that unites all three categories of the physical interplay (strong and weak nuclear forces and electromagnetism) with gravity, his insights spanned the Big Bang theory and imaginary time hypothesis. The union of the first three forces that rule in the microworld is fast approaching largely due to the hypothesis of quarks and quantum chromodynamics, while gravity has been shoved aside because it is too weak to influence the microworld. The problem that remains is the unity of the microcosm and macrocosm. This unity has never presented a problem for mystics and astrologers. However, for modern physicists, it is a problem to explain, prove mathematically, and test experimentally.

This problem is unsolvable because it is objectivized and projected from the outside, and the significance of the subject and the human personality in the whole affair is overlooked. The ultimate goal of objectivization of the outcome of nature is the Big Bang theory. According to this theory, the entire universe was contracted in one point of singularity some twenty billion years ago of an unimaginably great mass and temperature.

Since in that singularity (it is said "the size of a pea") all space and time were contracted, it goes without saying that it contradicts itself because it could not have existed anywhere and could have never occurred for one simple reason: There was no space and time before it. However, in the imagination of physicists, it was able to explode with ease and spread itself in an inflationary manner to create this existing universe. The proof for this singularity is found in an observation that galaxies continue to move farther apart, although the confusing bit remains as to their inability to determine the initial point from which they seem to be moving apart from - since it appears to be from everywhere. They are even more perplexed by the new unverified observations according to which innumerable galaxies do not seem to be randomly scattered across space but form a gigantic spiral, something like DNA. (This would confirm the Hermes saying that "nothing is outward, nothing is inward, for everything that is outward is inward.") The most distant photographs of the cosmos resemble the neuron images of our brains. Additionally, cosmic microwave background radiation, thought to be the evidence of the early stages of universe creation and a remnant of the Big Bang, surprisingly comes from everywhere.

The last drop that made the proverbial cup runneth over from which the theoretical physicists grow intoxicated was the result of the complete recording of microwave background radiation (Planck satellite 2013) that showed our planet is the center of universe. Apart from this, it proved that background radiation is coming from everywhere, dismissing the Big Bang theory, which would have to have some definite point of origin in space.

This brings us to the second cosmological theory, which states that cosmos is a big hologram, and its every part contains the image of the entire cosmos.[4] According to this theory, the results of background radiation would have been the same had they been obtained from any other observation point in the cosmos; each one would be at the center of the universe.

Those two theoretical postulates, the theory of the strong anthropic principle and the holographic universe, are the foundation astrology is based upon.

The universe exists with no beginning and no end in imaginary time, in an eternal present, which is sphere-like; it exists despite not having a beginning or an end. This meets the standards of both the theory of relativity

---

[4] On the hologram paradigm see Michael Talbot's book: *The Holographic Universe,* 1991.

and the quantum mechanics laws; they are free to interact now.

The concept of linear creation, from the beginning to the end, is a projection of a conditioned mind and every calculation for the survival of ego, whereas the non-linear existence, with no beginning and no end, eternal and whole (sphere and circle are the archetype symbols of human Selfhood), is an expression of reality of the being itself; the same as love, which has no calculated interest, it simply is.

The idea of existence that is without a defined beginning appears early on in religious and mystical speculations but was best hinted at by Plato in *Timaeus*, speaking of the spherical nature of the divine soul. In Indian philosophy, advaita *vedanta*, it is stated that the world is maya, an illusion, which has no beginning and no end. It will be clearer for us to comprehend the way how something that is without beginning can exist only if we perceive it as a sphere. The nature of the space–time universe, which is with no beginning and no end, is proven by the spherical shape of all celestial bodies and their orbits.

The four-dimensional space–time, the eternal present, which has no beginning and no end but simply exists, does not oppose the empirical, three-dimensional time that exists for our senses only. Its arrow is pointing in the direction of the future we do not know, while the past we partially recollect. In the eternal present, that past, present, and future exist simultaneously. In it the universe is infinite, unique, and without boundaries or singularity, and in our sensory time, things happen as though singularities exist, the universe expands, there are black holes, things originate and disintegrate from the horizon of our perception, and we believe that the new ones came into being or completely vanished while using the same logic to explain the functioning of the cosmos. In the eternal present, the "beginning" of the cosmos is like the earth's north pole, and if we equate latitude with time, we get the "end" as the south pole. However, these two points in themselves are nothing special. The same rules can be applied to them like to any other point on planet Earth. This analogy, however, shows that linear time exists for the subject only, who is on the surface of the sphere, in this case Earth.

If we add anthropic principle to this, which is not out of tune since we exist and envisage all these times, we will come to the theoretical explanation as to why astrological principles work to begin with. Firstly, astrology is based on the four-dimensional bending of the space–time continuum around the sun, which takes place by the spherical movement of the spherically shaped planets. Time in astrology moves in cycles (due to the spherical movement of nature) more in an imaginary way than linearly because it contracts according to the model of the hologram and is divided following the principle "one day, one year," based on which the solar horoscopes are made, primary and secondary directions. Every time unit has its own context. Therefore, ***the basic prediction methods in astrology are possible because of the characteristic of an eternal present due to the holographic nature of the universe and its superior relationship toward the three-dimensional sensory and linear time***. Everything exists only in an eternal present; the universe is a hologram. We can say that astrology is nothing but a method of translation of some basic facts of our existence, from the timeless, unconditioned present to the language and comprehending of everyday, limited, and conditioned sensory time.

The fact that astrology is based on the practical interpretation of the timeless holographic universe is best seen in examples when a certain planet is transiting a point where it used to be at the moment of our birth, it acts exactly the way astrological theory explains, and we are always able to see it in our personal experience.

Astrology functions according to the principle of holographic contraction of time, clearly shown in the effect primary and secondary directions have on our lives, where the movement of each planet forwarded one degree of the Zodiac circle - that is, one day forward - counts as one year.

All of this is adding more evidence to ***astrology as an ancient science that bases its practice on the holographic universe, which has been discovered by modern science recently, albeit theoretically***.[5]

Therefore, astrology is quite aptly named a science because in its essence, that is exactly what it is. ***Everything in astrology is based on facts***, such as the exact position of the planets, observing and analyzing facts, and their practical verification. ***There is not a single element of belief or superstition in astrology***. Besides, astrology has always been considered the mother of all sciences for a reason. For the proper understanding of astrology, what is required is conjoint knowledge of all the sciences without prejudice.

Planetary transits, which affect us regardless of time (and regardless of space because distance bears no significance), act on us as conscious subjects; it is the state of our consciousness and the versatility of events of

---

[5] Ancient knowledge offers a far bigger and richer perspective than the mainstream science would have us believe. More on that in the works by Graham Hancock.

our lives. In practice, this proves the theory of the strong anthropic principle.

This understanding becomes deeper once it is recognized in the unity of the overall nature. At that point, the spherical happening of eternal present of space–time, which astrology is based on, becomes crystal clear together with the fact that it unites cosmic events with the entire organic life and our destiny as well. From this perspective, namely, it is self-evident that Earth's orbit around its axis and the sun, as well as the remaining planets of the solar system, are directly responsible for generating all life on Earth and its dynamics.

With their gravity (magnetism), planets employing induction[6] generate even finer energy movements – for all intents and purposes, this is organic life. If Earth were to stop, together with all the other planets, all organic life would stop and disintegrate into dust. The only reason why we can set our psychophysical being in motion, more precisely, the reason why it can live, is because Earth, together with all the other planets, rotates, especially the moon. The moon is a gigantic magnet, and with its rotation, it induces most strongly the movement of the overall life on Earth. Its impact is the strongest because it is the nearest. The other planets also play their part according to size, speed, and proximity. Their collective influence through their mutual aspects creates life as we know it on Planet Earth.

Planets give characteristics and dynamics to our being depending on their dynamics. The moon is the fast-

---

[6] Let us remind ourselves here: When near a closed conductor without electricity, there is a moving magnet; electrical energy is generated that lasts for the duration of the rotation of the magnet in question. Faraday called it *magnetic induction*.

est, and it is responsible for the mind, which is ever-changing. Second in line is Mercury, which is responsible for the intellect, which is more stable than the mind but still has to be agile and quick.

The third one is Venus, responsible for love and feelings. It takes time for them to grow, but they are still apt to change and modification. The fourth one is Mars, in charge of energy and corporeality. It takes a lot more time to build the muscles and become fit, to acquire the strength and learn to apply it; our energy changes more slowly than the feeling of love and the mind. The fifth one is Jupiter, responsible for character. We cannot change our character and philosophy overnight. To change somebody, it takes years of assiduous effort, and even then, it does not happen too often. The sixth one is Saturn, responsible for wisdom, the slowest characteristic to develop in life. It may require a whole lifetime to develop a sense of wisdom, and once it is done, it cannot be undone.

The seventh one is the sun, which does not move at all. It remains fixed as the center of the whole system but moves through the galaxy together with all its planets. To other planets, it is immovable. The sun is the soul, completely motionless and permanent. For as long as there is the sun, the planets will be there too. When the sun burns out, all the planets in the solar system will be destroyed as well. Therefore, the sun is necessary for the survival of the planets. Mind, intelligence, wisdom, and love will come and go, but the soul remains solid, infinite, and unchanged as the witness of all happenings. Everything moves and happens around it and because of it.

The latest cosmology has not only taken one step closer to the ancient experience of astrology but has also confirmed similar hints written in the literary work of

Jains, members of the oldest religion, not only on the Indian subcontinent. In one of the oldest *Prakrit texts, Uttaradhyayana-sutram*, it is suggested that "time only exists in what we call specific space," and the classical commentaries explain that time exists only "on the continents where people live and the oceans belonging to those continents." Whereas in *Tattvarthadhigama-sutram*, it is quoted that "celestial bodies generate time through their motion." In the commentaries, it is further emphasized that time is also substance, and it is proved by the fact that both time and substance have identical characteristics: initiation, continuity, and cessation. Finally, it is stated that the function of time is to maintain the substances in continuity, modifications, movements, and their respective occurrences.

The latest trend of designing all of the existing insights into physical reality, advocated by Fritjof Capra, is systemic thinking. It stresses that intrinsic property of every living system originates from interactions and relationships between the factors where those relationships tend to repeat themselves in certain configurations or schemes. It is a dynamic process of relationships that expresses a certain quality and cannot be expressed quantitatively by measuring size and weight. Therefore, it cannot be expressed with one theory alone. Mapping schemes is what is required here.

It is hard to find a better definition of astrology than this. The natal chart is the scheme that maps our existence in time.

Systemic thinking shows us that looking for a single unifying theory is in vain because the microcosm and the macrocosm exist in a dynamic relationship that expresses itself in the form of a scheme through the natural cycles and phenomena, therefore through events themselves. The big unifying theory will be fully realized as our cultured living and not in some scientific journal that even the majority of scientists are not able to understand.

Astrology, as the oldest scheme of the phenomena of natural cycles taking place, defines our position in those events, the way they shape us and our living, the same way the celestial scheme shapes our beingness. By doing so, it definitely suspends the objectivation of the meaning of existence outwardly, which is nothing but the oblivion of existence itself. We are already in unity of the micro- and macrocosm and should not look any further through some objectivistic theory by means of which we would make some objective alternations to change our lives for the better. We should, instead, simply be aware of this unity because only such an unconditioned, pure consciousness can save us from conditionality. Only with it can we achieve something new and improved. The only thing conditioning us is unconsciousness.

Searching for some big unifying theory is based on the same conviction as the religious quest for the ideal faith that will "save the world," which further feeds the conviction that a collective "salvation" is possible. Every type of collectivism is just an expression of biological instinct for the preservation of community, herd, tribe, or nation, and the purpose of all collectivistic ideas (both religious and scientific) is the affirmation of such an instinct. The true "salvation," however, is an act of personal self-knowledge that is achieved through the process of individuation. Its orientation is, unlike the interests of the collective community, different; it could be said "vertical." Hence, there is no collective salvation. True freedom or spirituality can shine only through a mature human who will, with their personal example, testify of it to others - because it is the source of life and the human being. The essence is always uniquely manifested because it is alive. Only that which is alienated and dead, like some dogma or theory, remains always the same and can be distributed evenly to all. For the entire nature, its micro- and macrocosm intersect and unite in the human, not in some theory but in the integrated human personality.

Astrology displays the details of this unification..

♈ ♉ ♊ ♋ ♌ ♍ ZODIAC ♎ ♏ ♐ ♑ ♒ ♓

It is common knowledge that the word zodiac derives from the Greek word *zodion*, which means the "animal belt" and refers to the twelve constellations that encompass Earth along the ecliptic. It is less well known that this word comes from the ancient Egyptian word *zodjadja-kos*, which means "division for the sake of working." It is even less well known to what end the horoscope signs are positioned the way they are, in all four corners of the world and in all epochs, from North and South America, Egypt and Babylon, all the way to India, Tibet, and China.

The entirety of nature serves the purpose of manifesting divine consciousness through the conscious subject that is human-shaped on this planet. It practically proves that the entire nature serves to facilitate the enlightenment of the human soul. In order to make this happen, it must shape its energy, which is manifested in universal movement, coherently to be able to faithfully embody its purpose. It is this work of nature (i.e., cosmos) that is executed while shaping the conscious subject in miniature, in microcosm, and the purpose of existence of the subject in question is self-knowledge. Humanity is, therefore, a reflection of the cosmos in miniature, the microcosm itself, and this is the reason why astrological tradition, proved in practice, assigns an astrological sign to every functional part of the human body.

The head is shaped according to the principle of the sign of Aries, the neck belongs to Taurus, the shoulders and arms to Gemini, the upper part of the chest and heart to Cancer, the plexus to Leo, the stomach to Virgo, the hips to Libra, the sexual organs to Scorpio, the thighs to Sagittarius, the knees to Capricorn, the calves to Aquarius, and the feet to Pisces. Indeed, the human body is shaped the way it is because the earth is as far away from the sun and other planets as it is. If its gravity, determined by its mass, were any different, humanity would be different as well. Its mass and gravity are the way they are because it moves around the sun and other planets at the speed it does. And conversely, its speed is in proportion to its mass. If any of the factors were different, the remaining ones would also be different. Everything is interconnected and mutual. Separating the s from this wholeness is an expression of utter ignorance. He has been modified by the natural cycles, seasonal changes, and all elements that constitute nature, and it is not limited to the surface of the earth only. Its ground is just the surface where the overall beingness of the cosmos is projected on, especially the solar system because its mass, gravity, rotation, and revolution - the factors that the entire organic life are directly dependent on - are immediately conditioned by the planets of the solar system and their gravity and motion.

In the four-dimensional space–time, in reality, the shape, gravity, and movement of all the solar system's bodies, the organic life on Earth, and the human psychophysical body and destiny are one and the same thing but in the different dimensions and proportions. Their circular, more accurately spiral, movement shapes the different dimensions of phenomena. The grossest and the most objective dimension in our perception would be the very physical shape of the planets and their mass. They are directly dependent on the speed of movement and their proximity to the sun, which would, then, be the next, finer dimension. Weather and climate on those planets, which are also dependent on the previous dimensions, would be the next dimension, and organic life would constitute an even finer dimension than the one in question. Human character and consciousness expressed through destiny are the finest dimension of this whole spiral shaping induced by the sun. For this reason, the sun is the basic principle of integration of the human purpose. The sun, with its mass and gravity, bends and contracts the surrounding space–time, most concretely shaped as planets, and this contraction has generated organic life in the most suitable place of the solar system, on Planet Earth. Through evolution of organic life, all this contraction has realized its existential meaning. It attains its spiritual meaning through the personality of the most complete and most perfect organic being: man. That is why they are a microcosm in the exact same way as the astrology defines them.

The psychophysical evolution of space–time toward personality and the awareness of itself, in the favorable organic conditions on Planet Earth, is developed according to those principles found in the signs of Zodiac. They themselves are an expression of natural cycles and changes of the seasons. Each cycle of the change of seasons has twelve phases. Each phase represents one principle in the psychophysical beingness toward the consciousness, more precisely, one aspect of conscious beingness.

For the consciousness in the being to become completely mature, unconditioned, and defined, it cannot develop from one aspect of beingness only but rather from a series of aspects that constitute organic life, of which there are twelve. The experience of "beingness" must, by way of reincarnation, go through the whole natural cycle of transformation and growth before it reaches its Self, although it spins around it continuously. The sun has always represented the authentically human quality or essence. However, the nature of psychic experience is such that until it makes a full circle first, it is unable to determine where the center and the outcome of the whole affair are. It must face and confront everything it is not firstly until it finally becomes aware of what it is.

The cycle of a person's perfecting involves becoming aware that they have twelve phases, which are comprised of the four elements and three qualities each.

Twelve Zodiac signs can be divided into four groups of three signs each, representing the seasons. These groups best express the spiritual evolution, which is our point of interest here.

Pisces, Aries, and Taurus make up the first (spring) cluster of signs. It is a period of growth for the undifferentiated biological urges, which are expressed as instinctive, prerational character in man - everything that prevails from the sphere of unconscious and is impulsive, sensory, and imaginative.

Gemini, Cancer, and Leo make up the second (summer) cluster. Here the urges that have been previously set in motion stop for the first time with intent for things to get their place and shape, to differentiate and get sorted out. Humans, for the first time, separate from their undifferentiated natural origin and acquire the awareness of ego, themselves, and others and becomes familiar with things based on making a distinction between the subject and the object, but mostly on the level of physical shape (i.e., experience). Here things get sorted out and ascertained mentally (Gemini), emotionally (Cancer), and physically (Leo). By learning to reason, he experiences conflicting, disagreeing, and a clash of opposing views generated by rational distinction. More than anything, analytical reasoning and rational consciousness reach their maturing point here.

Virgo, Libra, and Scorpio are the fall group. The new evolution starts there, which, unlike the instinctive spring phase, is characterized by mental organization, harmony, and hierarchy, which will enable the spiritual ascension. At this stage, things are for the first time connected spiritually, according to their meaning and not by sheer reaction as before. A proper sense of balance between the oppositions, within the subject and in the objective world, between the I and the non-I, is born here. During the fall, the decay in nature seemingly erases life; but it is only temporary, clearing the way for new growth in the spring. The distinction between subjective and objective reality begins at this point.

The last triad is made up of Sagittarius, Capricorn, and Aquarius. Beingness has far surpassed the instinctive, rational, and egoic survival mode here and aspires toward the liberation from the space–time conditionality to the pure consciousness, spirit, and meaning. In the earlier stages, all experiences of the world of various phenomena and material limitations were being gathered. Now they are being assimilated and overcome in higher consciousness that is turned toward the Absolute.

Zodiac signs are divided according to their elements into the earth, water, fire, and air signs. The elements symbolically express ways of perceiving the world and reacting to phenomena. Those ways are thinking, perception, feeling, and intuition. These functions are in a mutually opposed relationship so that thinking is opposed to feeling, and perception is opposed to intuition. These psychological functions have their archetypal symbolics in the elements previously mentioned so that the earth corresponds with perception, water with feeling, fire with intuition, and air with thinking. This type of ontological structure correlates with the ontic structure of the world itself, which is comprised of the coherent states (earth) and fluid (water), which are dependent on different temperatures (fire), and gaseous or electromagnetic states (air).

The first triplicity of the fiery signs is made up of Aries, Leo, and Sagittarius. Through them, the intuitive, direct cognition of the world evolves from the unconscious impulsiveness of the Aries and its dominantly physical expressiveness on to Leo, where it is expressed as ego and authority, and finally with Sagittarius, where the intuition is expressed spiritually, uniting all knowledge and experience.

The second Earth triplicity is made up of Taurus, Virgo, and Capricorn. It demonstrates how the perception evolves from the one in Taurus, which is completely dependent on the concrete shape manifested and observable to the senses; to the one in Virgo, where perception is put through a series of tests where the result is that the sensory form is not the only one that is relevant; and the last one in Capricorn, where perception exceeds the sensory conditionality. The human soul in Taurus enters the material plane, in Virgo it begins to discover the true face of matter and goes through a transformation under the influence of spirituality, while in Capricorn the matter is crystallized and transcended to unconditioned spirituality into the consciousness of the soul. Taurus is ruled by matter, Capricorn is predestined to be the master of matter, and Virgo is a bridge in between the two.

The air triplicity of signs is made up of Gemini, Libra, and Aquarius. They show the process of evolution of mental maturity. Gemini represents the first impulse to learn; it aims to acquire as many experiences as possible, albeit more unconsciously than consciously, and it goes about collecting them without making much sense. Libra aims to harmonize all experience because it is aware of the fact that experiences themselves are nothing without the balance of opposing points because only a balance of that kind may produce the image of the whole, the purpose. Aquarius offers the acquired knowledge to everyone. It has gone past the stage of collecting impressions and their harmonization because it knows the meaning of experiences and is in a position to share it with others.

The water triplicity is made up of Cancer, Scorpio, and Pisces. Consciousness evolves through emotions in this triplicity. Initially, the first experiences are collected in Cancer, and they are child-like; they are experienced directly and absorbed in organized phenomena, mostly in regard to one's ego and body, therefore oneself. The emotional experiences are very personal there. In Scorpio, feelings are experienced more intensely and on a wider plane in relationships with others, which is done through intimate relations by understanding other people's subconscious feelings and drives and, in its crudest expression, through sexuality. The universe is experienced emotionally here by experiencing one's deepest impressions in a very profound way. Such experiencing happens in Pisces in a lot more liberated fashion; its feelings have surpassed personal limitations, namely, the need to see the world through the prism of personal experiences. They are open to the universal to the point that Pisces tends to get lost in it. To Cancer emotions are personal, to Scorpio they are connected with other people and the survival of interpersonal relationships, and to Pisces they are universal; they are dedicated to the infinite because in their feelings, there is no ego. For this reason, some think that Pisces is insensitive.

Zodiac signs best describe the evolution of the maturing of consciousness, which some religious ideologies interpret as the maturing of the soul to its divine stage. This maturing starts in Aries, where everything spontaneous aspires to manifest itself; it is the initial phase of energy charge. The sun seems to be the brightest at this time of year when everything comes to life. The next phase is expressed through an aspiration for the first concrete material shaping in Taurus, when plants start growing and first vegetation appears. Afterward, the first mental experiencing takes place in Gemini, which nature demonstrates with its changeable weather in the space of one day only, and the emotional Cancer, when all fruits get their final shape. After such a spontaneous and natural shaping, the ego is formed, the awareness of oneself based solely on the sensory experiences and personal feelings in Leo. Due to these personal limitations, the ego in Leo tends to impose on the people around them in a dramatic, even childish way, the same way the sun shines brightly in August and imposes its heat on the world. Once aspirations of this kind reach a critical point and acquire sufficient experiences, they start to be critically processed and analyzed in the sign of Virgo, and an alternative aspect of direct phenomena is sought after here. It is the time of year when fruits are harvested and earnings should check out. A spontaneous aspiration for analysis gets its harmonious outcome in Libra, a simple differentiation is exchanged for and aspiration to be whole, but on the level of ego, which means becoming whole in the relationship between I and others. This outer manifestation of wholeness and harmony can be seen in the objective reality in the world as lovely fall weather with its lavish colors. When this aspiration matures, a need for something wider and deeper appears: the need to be whole with the universal in the sign of Scorpio. The overall prior natural human evolution reaches its turning point here as it reaches its most powerful drive, which, if consciously prevailed by turning toward spiritual values, may be used as fuel to achieve higher dimensions of the otherworldly and spiritual. In the oldest astrological tradition this sign is dual as Scorpio and Eagle. The former one is a mundane creature ready to fight for survival at all costs. In order to avoid death in fire (of its own passions), it chooses initiation death of the egoic identification with the body and metamorphosis into an eagle that flies freely above the material world and the fire of natural urges. Tantrism is a faithful account of the metamorphosis so typical for Scorpio because in this sign, the living force is transformed through the ultimate temptation in serving the physical shaping; it changes direction and moves upward, and it becomes the driving force for spiritual uplifting. The experience of beingness touches rock bottom here, thus enabling itself to move upward. The time of year Scorpio occupies is connected with disintegration and decomposition of all foliage and vegetation in order for the new life to be born.

From Aries to Scorpio, an outward, purely sensory manifestation of the life force of nature was taking place: the sprouting, growth, and ripening of its fruits. In the time of Scorpio, it all comes to the point of disintegration so that the energy is transformed into the inner dimensions, and in the signs that follow the invisible, spiritual phenomena and values prevail. In Scorpio the soul attains the full spectrum of material and sensory observable manifestations of life (Scorpio is, therefore, the most aware of them.) and sets in motion toward the transcendence of those. If the crisis of this transformation is well executed, the person will become a true devotee of higher dimensions. If they fail, they will devote their life to demonism, which only superficially provides the illusion of overcoming materiality, whereas the material illusion is substituted for the astral one.

In Sagittarius spiritual experiences are united for the first time. Personal experiences of the previous phase, which were mostly emotional, open to objective values with the aid of abstract thinking and the systematization of all knowledge, often by means of philosophy or religion. The hope of some higher sense and new life is expressed in the time of year when the sun is in Sagittarius. After the gloomy, wet days of Scorpio, when everything falls apart, now the days are much nicer, sunny like the summer is coming again. In Sagittarius beingness for the first time transforms from the instinctive to that of the spiritual and human. The symbol of this is the centaur whose lower half is animal and the upper human. With the tightened bow, it aims in the direction of a higher goal. Sagittarians of the lower type are prone to expressing this initial experience of being purely human in a rather naive way through snobbery and insisting on titles and ranks.

Capricorn represents the maximum concentration of cognition and the experience of beingness, which resembles its season, in which all of the preceding forms have been reduced to their essence, to the seed in frozen ground, purified and reduced to its elementary form, waiting to start germinating and growing again. All processes of evolution, by contracting the space–time, reach their crystallization here, hence the only way from here leads to transcendence, to the unconditionality that enables the overall existence exactly the way it does, and to the spirit that enables the nature itself. A concentration of all psychophysical experiences (*samskara*) acquired in such a way, which further enables transcendence, is relayed onto the entirety of humankind through the phase of Aquarius.

The ultimate peak of reality is the insight that there are no others, that absolute beingness is absolute and without its dual counterpart, that our Self or essence is everything there is. If humans do not transform into their essence or Self, with all their maturity, they will maintain the illusion of duality, that there are other beings who need our help as though they are not the Absolute Self by definition already, as though anything could exist that is not the supreme and perfect reality. This illusion makes them give out their previously acquired cosmic knowledge to others through the phase of Aquarius. The reason why Capricorn does not attain transcendence lies in the fact that in their phase, the finalizing of the shaping of all phenomena is performed, and out of a tendency to hold on to all shapes, they decline into materialism and the need to establish their status and power over all aspects of forms - material, social, and spiritual. Aquarius, in a sense, corrects this mistake and sets itself free from the redundant forms. It achieves its purpose in the giving and the distribution of goods. More often than not, it does so in an intellectual and sometimes revolutionary way rather than the emotional way that it has long surpassed, so it appears cold, like the season it belongs to, windy and moody.

In the final phase, in Pisces, the universal values that have been mentally comprehended in Aquarius are emotionally contracted on the individual plane. The cosmic is experienced as personal here, and the universal becomes intrinsic, making this sign susceptible to all influences to the degree that it often loses itself in them, the same way that matter decomposes in frequent floods typical for this part of the year. If this disintegration does not lead man to the knowing of the unmanifested Absolute as the Self, then from the phase of Pisces, the soul has nowhere to go but to the next cycle of knowing the natural conditionality and growth. It does not, however, repeat itself in the identical manner: Reentering the signs the soul has already lived in means becoming acquainted with the higher spiral of their experiences. The karmic evolution unravels according to the dialectics of trying out the opposites even the richest imagination cannot begin to comprehend. It is exactly what we see as all living forms in all their shapes and contents. Often life in one sign is the counterbalance of the previous life. Life in the sign of Aries may prove to be a welcome change and will revive the direct contact with our living force and will after a restricting Virgo, and the life in Virgo marks the return to the concrete after the experience with the abstract in Aquarius. Therefore, we do not live in each sign only once. With each new entry into a certain sign, we experience its finer vibrations and meanings. That is the reason why we see very different individuals belonging to the same sign, the "higher" and the "lower" types, from the crudest to the most spiritual ones.

  It would not be wise to conclude from this review that only a Capricorn is "predestined" to achieve spiritual enlightenment, although this sign is at the top of the list of great devotees. The dialectics of nature presented here - the mutuality of all of its principles - proves that no sign is able to exist without all others. The rule that applies here is "all for one and one for all." Only when all signs join together can they make up the whole man. We can add that each sign has twelve dimensions, but it would be too much to start a debate here. If Capricorn enables the highest speed for the transcendence of natural causality, this property has been enabled by the characteristics of all other signs. Actually, each sign has its own path to liberation and spirituality, its vision and means of how to turn its nature into the deity, and how to recognize divine consciousness in nature. Each sign is only one out of twelve aspects of nature that is unique.

As a result of this, throughout history, several ways have been devised for the knowledge of the divine. Two main streams have been put to the fore among the many others: the path of submission and the path of self-realization. The path of submission corresponds more with the "lower" signs, from Pisces to Libra. They go directly through experiencing various aspects of natural urges, which suit them for reaching awareness. It is the tantric way that advocates that the entirety of existence is the reflection of the divine perfection and wholeness and can, therefore, be used for the purpose of self-knowing. It is a path of conscious acceptance of any form of natural manifestation: from sexuality to astral visions of deities and the idea of God. The other stream is the path of self-knowledge through detaching the human soul from natural necessities. He goes against the stream of natural causality into transcendence, and he finds human authenticity in the spiritual conquests of the unconscious natural elementals. This way is suitable for the "upper" signs, from Scorpio to Aquarius. Both streams lead to the same goal and do not exclude one another but rather complement each other. They simply show that one may reach the unity in many ways, depending on the character of each individual.

☉ PLANETS ☽
☿ ♀ ♂ OR ♃ ♄ ⛢
THE PSYCHODYNAMICS OF SPACE-TIME
♆ ♀

The influence of planets on the life of people has always been a matter of mystifications. In the early phase of its development, astrology was under the influence of magic thinking, therefore the understanding of planetary influence was demonic. It was thought that every planet has its own spirits, which the magician, meaning astrologer, may summon to learn all about its influence. In those days, that was the only way to find out about the impact planets have on our lives, and it was not a figment of imagination or delusion. Nature in those days used imagination to cultivate the people. When a magician summoned a certain spirit, this imaginary creation from this astral realm would really convey information to the questions that were posed. However, those answers were usually in accordance with the demonic belief of the practitioner because astral is imagination of nature, which, like a mirror, reflects the convictions of those that approach it. Therefore, no nymph from Venus was able to inform her astrologer-magician of the facts of existing planets, which we will delve into here; instead, she made an effort to seduce him with her dazzling beauty. Besides, maybe every planet has its own spirits if Earth itself is laden with them, but they can only testify of the special characteristics of their planet, not about the fundamental issues of all planetary influences.

The answer about the essence is available today based on the knowledge of the nature of physical reality we have disclosed so far. Therefore, the physics itself will provide us with a quality answer to the question of what the basic influence of astrology rests on: the influence of planets.

Our planet is a gigantic induction machine; the earth is zero, and the ionosphere is the phase. Progressively, with every meter up in the air and away from the surface of the earth, the voltage increases by 100V, creating an enormous energy. This energy is not only electric but it is shaped into all forms of living energies we see as the organic world; it is bioenergy that modifies and moves the overall life on this planet. Movement is no different from the energy and life itself, which is the manifestation of energy. The word energeia means being in motion, in perpetual movement. ***The entirety of life on Earth receives its movement energy from Earth's rotation and revolution. Therefore, we move and live only because Earth is constantly in motion; its every movement enables the functioning of our hands and heart. If Earth stopped, our hearts would stop.***

If the earth rotated all by itself around the sun, the manifestation of the energy induced in this way would not be complex enough to shape the life we have today. It takes more movement round the earth, which will, with its magnetic induction, differentiate life on it. Such induction machines are represented by the planets rotating in the solar system. Why there are so many of them and why they have the mass they do, as well as the gravity and distance from Earth that sparked the evolution of organic life on it, is best answered through the strong anthropic principle. It was all needed in order to shape the conscious subject. The natural constants are such.

Each planet, with its gravity and movement, induces a certain differentiation of bioenergy on Earth. The sum of all planetary influences shaped all life we see following the same pattern, together with shaping our destiny as well. It is hard to tell them all apart since it takes an enlightened mind to perform this task, but we can detect some of them. They are the easiest to see in the cycles, especially those of the moon and Saturn, which are associated with emotional and mental maturing. It is easy to identify a single-year cycle, when Earth makes a full circle round the sun. Based on this cycle, the yearly horoscope is made, or the chart of the solar return.

Medical science discovered that human skin regenerates completely in the space of 28 days. By the time the moon makes a full circle around Earth, we get completely new skin. Maybe the same science will discover that the entire body of man transforms and renews all of its cells in a year, from one birthday to another. If it is focused enough, it will discover that regeneration is well regulated by the time a person is twenty-eight years of age (somewhat faster in infancy), and from that point onward, it begins to slow down. The cycle of Saturn, interestingly enough, is of identical duration.

What are planets actually?

They come about in the same manner as everything else, from the universal quantum field. The quantum field was known earlier as ether (or *akasha*). In more recent times, its effect in nature has been called "dark matter" and "zero-point energy" as well as "black holes." Black holes are merely fields of blank space or ether. However, in mainstream science, the importance of ether is marginalized,[7] and its impact, which is impossible to conceal, is given these various terms.

Everything originates from ether/akasha or the quantum field.

Ether in the form of black holes is abundant. There are very tiny ones and those that are undeniably huge. Their sheer size creates the proportions of the phenomena manifestation of everything.[8]

Proportions range from the absolute, which enables the timeless space that in turn enables everything else; to galaxies, which are next in line; all the stars, including our sun; planetary systems around the sun; and the earth and the entirety of organic life on it. Following the trail of this logic, the biggest black holes create galaxies, smaller ones create stars, even smaller ones create planets, more minute ones create conscious living beings, and the tiniest ones create electrons, which form atoms together with all elements.

Black holes create all of that according to the model of torus.

Everything that exists in the physical universe, from the electron to the man and stars, even galaxies, exists according to the model of torus.

---

[7] The notion of ether is forbidden in science because Nikola Tesla discovered that limitless energy could be harvested from ether free of charge. Ether was thrown out of the Periodic Table of Elements by D.I. Mendeleev, where it occupied the most prominent place as the earth's ether. It was thrown out of J.C. Maxwell's equations that dealt with Hertzian and non-Hertzian waves, the manifested and unmanifested nature (i.e., ether). After the death of Maxwell, the scalar part (longitudinal electrodynamic waves) in the equations was left out, and only the vector potential remained.

[8] On the proportions of the manifestation of the cosmos see the book by P.D. Ouspensky, *The Cosmological Lectures*.

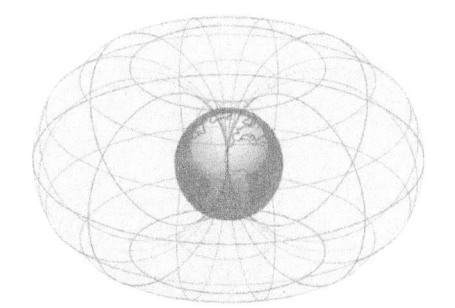

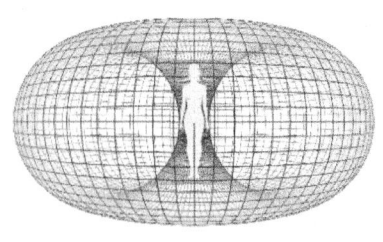

At the center of every torus there is a black hole, empty space, akasha, or ether.

Their varying sizes produce every difference in the shapes of everything that exists in the cosmos. The biggest black holes generate galaxies. The smaller ones make the stars, but their magnitude is such that in accordance with their proportion, on the verge of their torus, they induce a reversal field (generated in interaction with surrounding space), which creates intense vibrations of all elements, and due to the intense vibrations, extreme temperature is created as a result. It is manifested as the hot plasma we see on the surface of the stars as their light. The interior of the stars is cold and dark due to this kind of structure and ether in the center.

When the black hole is smaller, due to the nature of its proportion, on the verge of its torus field, it creates hot plasma made up of all the elements, but apart from plasma, it keeps the elements in a cooled down state. This happens because the black hole is smaller than the one in stars and the elements are able to form in cooler conditions, not only as plasma but as the concrete elements we see as planets.

Planets and all celestial bodies are spherical in shape because they originated from the toroidal movement of energy.

Therefore, every planet is hollow in the center with a small star inside, which in turn has a small "black hole" within. Planets keep the vibrations of the inner sun in their crust because the crust has not been evenly cooled, and underneath the cooled crust, there is magma, which sometimes erupts in the form of a volcano. This phenomenon happens according to the pattern of torus also, and the volcanos on Planet Earth and other planets occupy the identical latitude. However, not all planets are of the same size. and the bigger ones, with a little bigger black hole in the center, have a higher temperature that disables the forming of the hard, cooled crust made up of elements. They therefore exist as gas giants (Jupiter, Saturn, and Neptune).

Human beings and planets originated in the same way, from torus. In the gravitational center of our being, three centimeters (about an inch) below the navel, every man has a black hole, a field of pure ether or *akasha* (it is *Tan Tien* in *Qi Gong*). **That is why the planets and people align in energy synchronization and synergy. Synergy is the foundation of the effect planets have on people.**

Since this is a process of the universe itself, human characteristics and dramatics of functioning (destiny) cannot be determined by the subjects themselves, because they are the ones that are being determined. Still, this is not as fatalistic as it sounds. If the universe created a subject, then the subject's essence, or Self, is the universe itself, more accurately, the unconditioned divine consciousness that enables the universe into existence to begin with. This is, therefore, a very optimistic claim. It sounds so positive that all theologists and their followers, led by politicians, unite against it. All so-called "rational" people would join in together with the scientists because their mind is not programmed for such broad visions on the meaning of the existence of man. They are accustomed to more modest definitions. In accordance with the existing mind frame, imposed by religious and scientific education, it is not hard to envisage a soul that transcends the body and have faith in God that creates the world, but to even think that what enables God and the overall cosmic creation is our true essence; that is unimaginable. All secret controllers of polity and societies on the whole will stand up unanimously in their condemnation of such concepts because everything must remain within the set boundaries of common sense in order to be handled with ease. This is how it should be since the idea of freedom as the essence of everything that exists, ourselves included, cannot be served from the outside; it must manifest of its own accord as individual maturity. That is the only thing it can be. Until the human is fully mature and answerable for their actions, society must keep a tight lid over them and control them within the boundaries of their own mind; otherwise, the idea of absolute freedom may open doors for a hellish abuse. Freedom will happen but in a more distant future.

The evolution of the conscious subject is determined the moment he shows up as an individual in the three-dimensional world. More accurately, it happens when their physical body is born. It is the making of the physical world, and it is quite understandable that all human activities will be conditioned by the physical laws, both biological and cosmic, that made the biological ones.

We have concluded already that the human body is the way it is because the solar system is of this magnitude, gravity, and the movement of all of its objects. Humans would look completely different if the mass and, automatically, the gravity and movement of any other planet were any different.

Human appearance and character have been determined with such precision for the same cosmic reasons their actions were determined as well, because one person's character and actions are not that different. Due to the egoic and sensory observation, we are under the impression that physical appearance is one thing and the destiny of their movement something quite different. It is a relatively naive misconception, deeply rooted in all those who "do not believe in astrology" because they fail to see something that is painfully obvious, namely, the body itself is nothing but the matrix or means toward a certain goal; its shape is highly functional, but it has not been created for an empty purpose to serve itself as an object or to pose but to be an instrument one works with. The purpose of the body is, therefore, beingness-in-time and not existence for the sake of existence. For that reason, cosmical and biological laws that shaped the body into one organic whole, shaping its movement and activities in the world. There is not a fundamental difference in the causality of beingness that functions within our bodies, and the connectedness of our organs into one living whole, outside of our bodies and in our movements and actions in the world. For the unity of nature, our skin is no boundary; nothing is only outer or inner to it. Being an active participant in life by traversing the globe and experiencing life puts us in the same organic unity with nature as the movement of atoms and molecules in our bodies are connected to the movement of the planets, stars, and galaxies.

Everything that originates at a specific time is determined by the zeitgeist of the epoch because time is no different from the shape and space that are modified by cosmic objects. The space–time is spherical. It means that time is determined by spatial modifications; it does not exist all by itself, and it is not some abstract category. It is rather a matrix of spatial phenomena. Therefore, what we do with our bodies throughout our lives will be determined by spatial factors, where the planets play the biggest and the most powerful part.

The body the mind belongs to is preplanned for the beingness-in-time, which is only possible in the three-dimensional conditions of the organic realm where time is linear (past-present-future). It applies to the surface of this planet only. If we were in the center of it, we would be unaware of time to the same degree we are aware of it now. Our clocks that show chronological time (by the way, the farther away from the surface of the earth they are, the slower they tick), are built according to the model of zodiac division of time cycles into twelve equal parts. Organic life cannot exist in any other way but in the three-dimensional linear timeline of events. This should be clearly understood if our intention were to grasp our destiny as being preordained by the motion of the planets and the position of Planet Earth and no different from the organic life that has originated by this very motion.

The planets have a full impact on our lives and define our actions through their positions and the aspects they make.

All types of actions have their equivalent in the houses of the horoscope. There are twelve of them, the same as the zodiac signs, and their meaning is similar to those of the zodiacal ones.

The shaping of our destiny in time and space is shown in astrology by means of houses, planetary positions, and the aspects they make. The natal chart is the zodiac circle divided into twelve houses. Division is determined by the ascendant, which is the line of horizon observed from the place of birth (latitude and longitude) at the moment of birth. For example, if someone were born around midnight, at the time when the sun is in Aries, the ascendent will be in Capricorn and the sun in the fourth house. Therefore, their sun is right underneath the ground we stand on. To those whose birth occurred at noon, the sun will be in the highest position in the skies, above the head, and ascendant will occupy the sign of Cancer. When someone is born at daybreak, ascendant will be the conjunct sun in Aries also. Therefore, the time of day (the hour and the minute of birth) is the crucial component for the placement of the twelve zodiacal houses, which, combined with the signs and planets, completely determine the destiny of the beingness-in-time of our body. The meaning of the houses is closely linked to the meaning of the twelve houses in question. The first house bears maximum resemblance with the sign of Aries, the second one with Taurus, and so forth.

Houses represent the goal of existence and orientation in space–time, otherwise known as destiny. The goal of existence is determined by the position of the earth and the sun's gravity together with the other planets and their aspects. The gravitational pull of the sun is by far the most powerful; therefore, its position regarding the place of birth, the house which it occupies, has the deciding influence on the destiny of an individual. The destiny of those born at midnight will be to spend most of their life at home and follow a certain tradition and family matters; they will generally be preoccupied with introspective thoughts because this is generally associated with the fourth house symbolics, as if the sun's gravitational pull, which is underneath the ground at midnight, keeps them glued to the place of birth. Those who were born at noon experience life that perpetually draws them out in public; they are always career oriented and traveling to remote places and exploring foreign climes. The sun's gravity at noon seems to pull them upward, where they are always exposed to the public eye, and they like to be seen in public and large, open spaces. The one who is born when the sun is rising will have a strong personality and self-consciousness because it is the meaning of the first house, where the sun is placed at that hour; it is dawn, and everything shows their true colors then. The characteristic of people born at sunset (the sun in the seventh house) will be a prominent social life and dependency on the environment because it is the time of day when people process impressions of the day gone by, often in the company of others. A human born in the early morning will have the sun in the twelfth house and will spend the larger part of their life alone, suffering the influence of their unconscious resting, like in sleep. Leading an isolated and withdrawn life is one of the main characteristics of the twelfth house. Those who were born in the evening, right after the sun sets, will be filled with daily impressions for them to analyze and systematize, with a marked collective consciousness, which is characteristic of the sixth house. Those who were born in the morning, between the hours of 9 a.m. and 11a.m., will have the sun in the eleventh house and will be prone to gathering their own social groups, whereas those born between 8 p.m. and 10 p.m. will have the sun in the fifth house and certain inclining toward individual expression, frequenting places that can provide them with personal pleasures. They will do what they like and brag about the ones they like. Around midnight, everyone's at home, and at noon, they are all outdoors somewhere. Between 12 p.m. and 2 p.m., the most mature part of the day sets in, when the peak of wakefulness and mental activity serving common benefit to the society reach their highest point, which is the meaning of the ninth house, where the sun is placed for those born at this time of day. The sun in the third house will be found in such individuals who were born between the hours of 12 a.m. and 2 a.m., and in this instance, human consciousness will be so preoccupied with personal contents that it will be oblivious to everything else in life because these are the hours when people sleep. When, between 2 p.m. and 4 p.m., the sun begins to wane, we are aware of the fact that a day is drawing to a close and, with it, the significance of all this has on our life and everything we receive as our daily bread, which best describes the symbolics of the eighth house. Between 2 a.m. and 4 a.m., each of us, with the first occasional slips into our waking state, begins to develop awareness of themselves in the physical body, which is the meaning of the second house. This layout has been oversimplified and serves only to demonstrate how the position of the earth structures our psychodynamics in space. Our true destiny is determined by all factors of our horoscope, all planets and their aspects with one another, and some other points of influence.

While the signs are static and represent the global frame of phenomena, planets are dynamic factors that introduce the dialectical laws and the dramatics of events. Apart from the placement of planets in houses and signs, their mutual aspects are the chief factor of shaping our beingness-in-time, more accurately, our destiny. Each planet has its own speed and proximity so that the mutual aspects can never be repeated. If it so happens that they are eerily similar, which may occur in the space of over a thousand years, the earth will definitely not be in the same rotational position; therefore the houses will vary. For that reason, there are not two identical lives in this world. Every experience of existence is different because everything is in motion.

Space–time is functionally determined according to the laws of geometry and numbers. They are used to demonstrate the determination of space–time created by the planets with their mutual relationships. The zodiacal circle is an accurate depiction of the wholeness of space–time, and its creative dynamics are manifested by the even division of this whole. All divisions of zodiacal circle (360 degrees) into two, three, four, five, six, seven, eight, or nine equal parts make one effective aspect. The division into two, three, and four parts bears the most significance. A division into two creates the opposition aspect between these two planets, their direct confrontation, which, depending on their nature, may be the cause of a conflict between their influence, a constructive differentiation, or a complete breakup so that at times, one planet is in charge, and at other times, the other one takes over, according to the model of seesaw. Division into three gives triangular shape, which is an ancient symbol of creation. It is the most favorable shape because the two planets unite and give their best without engaging in extreme conduct. Division into four produces a cross, which is the hardest aspect to handle because the influence of the two planets crosses like swords as they cut one another. Number four also represents a square, which signifies realization or materialization, which shows a tendency on the part of both planets to have their way. Both of them would like to get realized to the same degree, hence the conflict of interest, where often enough, one overpowers the other, creating pressure that is responsible for the destructive elements in the character. That is why the aspect of the square is the most influential in horoscope.

The nature of the planets is such that each one represents certain psychophysical functions. In order to understand them, we should first understand that the psychophysical functions are effectively ways of integrating space–time into consciousness; they are the dialectical principles of integrating cosmos into the conscious subject. We have seen that this integration happens by contracting the space–time as a result of the motion of Earth and the other planets. In this way, psychophysical functions are determined by the planets and their position in relation to Earth. The key role of each planet is its distance and the speed of movement. They are divided into categories into the personal (those from the sun to Mars) and super-personal ones (Jupiter, Saturn, Uranus, Neptune, and Pluto).

Planets that are closer to the sun are easier to manifest in our daily functioning, in communication (Mercury), a sense of impeccable conduct (Venus), emotions and moods (the moon), and investing energy into all that (Mars). The more distant the planet is, the more profound its influence on the character is; therefore, Pluto is indicative of the deepest self-consciousness. Its orbit (together with Neptune's) is the most distant, encompassing the orbits of all other planets and inducing the phenomena of psychophysical functions they represent. It appears that the sun, aided by Pluto, determines the motion of the whole system. While the sun naturally manifests in people through the zodiac sign and aspects, Pluto is the pinnacle of our conscious accomplishment we have achieved through our work by overcoming natural manifestation. The house it occupies will show us how we can express our most profound values.

Planets, with their gravity, attract orientation of our beingness-in-time, and each one, in accordance with its characteristics, pulls us toward itself. It sometimes happens that more than one planet can be found in one sign or house (*stellium*), which is easy to recognize in a person since it causes a certain one-sidedness in behavior and the aspirations in life, depending on the sign and the house in question. Sometimes they are evenly distributed all along the zodiacal circle, inducing energy to flow in many directions, and the person is not quite sure what he or she wants. Most often, though, there are configurations and aspects in horoscopes that determine the orientation in life, and they can be interpreted from the natal chart successfully. Planets in signs, houses, and mutual aspects, ascendant and house rulers, all together make up ninety percent of all chart interpretations. The only reason for the wrong interpretation is the inability to recognize the true nature of each of these influences and not being able to tell them apart. In that case, certain natal influences are mystified or wrongly emphasized, and the solution is sought after in "past lives." Indeed, for those who do not wish to see their life, they always look for the answers to its many challenges elsewhere and not where they should.[9]

The sun is by far the most influential factor of the entire horoscope because it possesses the strongest gravity. It is the pivot of the overall motion; everything rotates around it, and all aspects of the other planets are directly dependent on the sun. It is obvious that the entire horoscope depends on the sun and the entire living world as well, and it is therefore not a very serious statement made by some astrologers that its influence could be reduced down to a small percentage at the most, because they ar-

---

[9] To become acquainted with past life influences or parallel lives there is karmic astrology, but it also is just another way to interpret the contents of the already existing natal chart. To become acquainted with other incarnations, hypnotic regression method is more effective.

rived at this figure by simple division of all the factors that constitute the chart.

The active influence the sun has, as well as all other planets, depends on the aspects it makes. If there are no aspects, then its direct influence has been disabled, leaving the planet with only passive functioning, combined with the influence of the house and the sign. For this reason, the number of aspects in the chart is the best indicator of the effectiveness of an individual; any aspect is better than no aspect at all.

The problem with the sun, and the reason why some people believe it is not the most influential, is the problem of properly grasping the nature of its influence. In order to understand it, one should first realize that the sun does not move like the planets, following a set path, although it looks that way from our point of observation, but everything rotates round it. The sun is static. It is the outcome of all phenomena and not merely one of the factors. When it comes to shaping a subject and a personality, it could be said that the sun is not merely one of the psychic functions but the center around which all of the psychic functions are shaped by planets. On the earth, only our bodies are shaped. The planetary influences shape their entire psychodynamics or ego, temperament, and character, and all this revolves round the sun. This means that the sun is the outcome and the chief attractor of all phenomena; it is the center of the psychophysical beingness, the consciousness of our soul or the Self. That is why the Self (sun) is independent from the body (Earth) and the whole psychodynamics (planets), the same way our soul is independent from everything.

Therefore, it should not come as a surprise that its influence is not so apparent in a vast majority of people. It is much easier to recognize the influence of the moon or

Mars, for that matter. The influence of the sun is not manifested directly but indirectly via the moon and other planets because it never acts alone but through a whole system.

Depending on the sign it occupies, the sun demonstrates the karmic maturity for the Self or consciousness of the soul to integrate completely and manifest through the body of man as the human's awareness of their own existence. Each individual sign is one way of the Self manifesting itself, an expression of the soul. Since Self is the essence of all beingness, it is not static but dynamic and cannot be manifested in one way only but in twelve, the same number the natural cycle of the overall transformation and creation of life is divided into.

DESTINY RESIDES IN THE MATTER

Destiny is a process of harmonization of life in order to make it suitable for the absolute reality that enables everything, reality that is the essence of everything that exists. For this very reason, destiny is a natural process. Everything we do our entire lives is essentially nothing but meeting the needs of our organic existence and following our destiny. We are subjugated by our destiny until the point we become capable of receiving the self-knowledge of our transcendental soul. ***Once we achieve that, we overcome our destiny. Our soul is above the organic life, space, and time. Destiny oppresses us if we fail to do anything for our spiritual outcome, and this lack of results is a key feature of all bad destinies.*** Our dissatisfaction with life is a direct consequence of us failing to contribute to our spiritual authenticity, or we do it in the wrong way because there is not sufficient information of what it really is. Numerous traditions present it in a warped way.

In antiquity, the finest natural processes were detected in microcosmic dimensions, and they were called matter. This term appears in Aristotle's teachings as a name for substratum, the matter something is made of. The very word was used primarily to denote wood, forest, or building materials in Greek. Later, Aristotle introduces the term into philosophy as that which lies at the base of all matter, the ground for the existence of the attributes of substances, that which can be modeled and acts as the carrier of all properties. He considered it to be proximate matter. The Greek term for matter literally translates to *materia* in Latin. Modern physics has found these finest natural processes to be not such simple creations. Subatomic research studies have shown that matter is, in effect, an energy creation; what atoms or substance consist of is in reality an energy unit named a quantum field. It has been discovered and proven that nature on its finest level is an energetic phenomenon that is directly dependent on observation, even the presence of the very subject. This has been mentioned before and goes to prove a simple viewpoint that the overall energy motion of nature aspires to shape the subject or is, at least, dependent on it.

In a much older experience, older than not only modern physics but antiquity as well, in the Jain religion, the primary substance of nature is called *jiva*, which means life. It is interpreted as the living energy creations that by way of their complexing generate all elements available to sensory perception together with all phenomena. According to the Jain religion - which is the oldest because it goes back millions of years, although its reformers and bringers are well known and date back 4,500 years; its founder still remains a mystery - the primary substance of nature are the energy monads that are alive (*jiva*), which means they are conscious and therefore functional. In their development and complexing, these monads can be differentiated into those that are in charge of the organic, sensory, and visible living world and the ones that form the inorganic world, which in the sensory experience seems to be inanimate. In both of these cases, they are functional and, through their action, generate matter. In Jain belief, matter is "sticky substance"" or the "excretion" of the functioning of *jiva*.

The Sanskrit word for this function is *karma*. Later speculation in Indian philosophy and religion on the notion of *karma* added quite an altered meaning to it: the punishment or reward for the actions done. Initially, it denoted the functioning of the being itself and the functioning of the conditioned nature, and Jains consequently aspired to attain independence from it (*kaivalya*) through the ascetic purification of their souls from all the filth of nature conditioned by actions, from *karma*, which is like a sticky substance glued to the soul. With this aspiration to be the winners over the whole conditioning of the beingness, they were adequately named Jains, which means "victors." They belong to the hero cult, which is much older and opposed to the cult of gods. When humans triumph over the conditionality of nature, they discover that their soul is not a part of overall existence but rather belongs to the unconditioned divine consciousness, which enables existence: its essence, soul, or the Self.

We have seen how the complexity of the motion of these primary energy monads on the macrocosmic plane takes place by means of bending space–time into forming the celestial bodies and organic life, something astrology testifies about. This motion we can now recognize in the form of life experiences or destiny because there are not that many differences between the two; it is all one and the same natural process. It appears different only because it plays out in different dimensions. The issue of destiny and karma is rudimentary for us to be able to understand how astrology works, and it is therefore necessary to distinguish the meaning of these words from the bunch of "excretions" and "sticky substances" they are glued to.

The whole point of *karma* is action but not our action; rather it is the action of nature in general. In *Bhagavad Gita*, it was written that the purpose of liberation is freedom from karma, which is achieved through nonattachment as regards the actions; humans should not attribute to themselves the activities, not even the psychophysical ones, because thoughts are not our own either. It all exists and acts in any way possible as the activity of nature, and not human essence, the soul, which is the eyewitness (*Sakshi*) and has the attributes of pure consciousness and unconditionality. Not only does the entire cosmos exist in order to shape the conscious subject but it also exists to lead this conscious subject to the highest enlightenment. The spiritual essence if humans is, therefore, the goal of all phenomena, not the other way around. For this reason, **humans should not be the slave to natural phenomena seeking sanctuary within, because all phenomena seek its outcome and purpose in the human soul, which is manifested in this world as personality**. The more integrated the personality is, the less identified man is with the objects. All paths of human salvation are reflected in the ability to distinguish between these discrepancies and the cessation of identification of the eyewitness with the phenomena. It is the road of transcendence. The eyewitness, then, becomes the master of phenomena; they are not at the mercy of it like before, when they were unaware of their true, transcendental, and unconditioned spiritual purpose. They no longer have a destiny, because they have stepped outside of time.

Therefore, the only way to overcome destiny is to understand that it is a natural process and not the phenomena of our essence, Self, or the soul. It is primordial unconditionality that enables the being to be, making our Self at the same time the Self of all beings and of existence itself. Consequently, it is never threatened by any phenomena or destiny. However, the average person finds it hard to accept this as reality. They are convinced that destiny keeps happening to them, saving or jeopardizing their life. All destinies exist primarily due to this faulty conviction.

Astrology does not exist as a method of divination, but it exists to enable the understanding of the true nature of everything that goes on. It shows the way in which the human essence or soul is glued to **natural activity**, which man experiences as their destiny only to the degree they are identified with what they are glued to. Humans, in their ignorance, blame God for their "misfortune" - or the devil, most commonly other people, and sometimes even themselves. By doing so, they only add to the number of conflicts. They do not decrease over time, because they are profoundly unaware of the cause of the whole affair. If they knew the laws and influence of astrology, they would see that it is all an impersonal natural process he can *differentiate* themselves from - all more clearly if he grows to understand the direction in which astrology is pointing.

In our natal chart, we can clearly see the plan of our lives. Problematic aspects signify temptations or tendencies we must overcome within ourselves, that which binds and enslaves us unconsciously and the traits of our character we have not adopted well yet and learned to implement harmoniously. Good aspects point to the assistance and support for our growth we should learn to use in overcoming the difficulties. The natal chart shows what the main topic of this life is and all accompanying experiences, what our task is, and things we should work on in order to make headway toward perfection.

Planetary influences determine not only the character of an individual but interpersonal relations as well. Compatibility charts (synastry) will indicate to anyone who wishes to double check experimentally, that all details in relationships between the two people, or one man as regards everybody else, are thoroughly conditioned by certain planetary influences. All phases of the relationships are evident, the times of crises and conflicts, too. Relationships where one person's self-interest prevails are very common, as well, sometimes temporarily and sometimes forever, without the other person realizing it. It happens when someone's destiny is so strong that it pulls toward itself the life of another person with such powerful gravity - the one who is predisposed to serve this purpose with the corresponding planetary positions, to act as their assistant, a go-between that conveys certain information (sometimes it is just a good book), to be the physical laborer, and, at other times, a victim too. The destiny of many people is to be someone's servant, even victim, and this ugly reality they do not want to see; they do not want to hear about this astrology because it reveals the truth about themselves. Many cases could be found in which a very destructive astrological character of one person or an object (this was rumored for the Titanic) sinks and drowns everybody they had their hold on. This is the most apparent in a family situation, where due to the very destructive character of one family member, the entire household is doomed. The same applies to the positive influences. One grand and advanced soul enables the whole community to reach considerable progress. Much like everything in nature, destinies of people are mutual, not isolated, processes. All bodies in nature have a gravity of their own, and the same applies to people. Once a baby is born, it disrupts relationships between the existing family members and introduces a whole new lifestyle - the same as the new celestial body entering a planetary system and altering the pre-existing trajectories. Gravity has its own system of subordination, its higher and lower, grosser and finer, manifestations. The grosser manifestation of gravity is what we perceive to be the motion of celestial (and other) bodies, while the finest manifestations of gravity are the psychodynamics of an individual and interpersonal relationships, in other words, human destiny.

Aside from the individual *karma*, there is the collective one, representative of the whole nation. Often it is stronger than the individual one and tends to prevail. It is something to think about if we truly want to understand destiny and *karma*.

As far as the relationships with other people and the outer world are concerned, the human finds it hard to believe that those events are a product of the natural processes indicated by astrology. They think they are theirs or they were caused by "others" because they lack sufficient awareness of what the relationships and phenomena are to begin with. When their health or physical appearance is at stake, they are then a lot closer to the insight that their destiny has been predestined by something outside of their conscious scope, some "higher cause"; it "happens" to them, and they are in a way different from it. It seems that only a hard and unfortunate destiny is sure to lead to the intuitive insight that our soul is not dependent on the physical body and other impersonal natural phenomena that have a tendency to mercilessly destroy any physical shape. In nature, the death of a physical body is so simple and easy to accomplish because our bodies are not our true essence; they are merely a secondary, outer form, a layer over the soul.

more accurately put, the organic creation in which the soul is allowed to realize its three-dimensional presence as the most concrete form of presence. No other dimension offers the essence of existence to our souls nor does it provide it with such concrete manifestation as the three-dimensional organic world does through the human shape, body, and impact. All remaining dimensions manifest the essence of everything through the impersonal existence.

Astrology practically and theoretically displays all the details of the "higher force" that governs the body, but in order to be ready and willing to reach out for the stars, we must first understand the regularities of karmic evolution.

Its starting point is the primary law, which states that the universe exists with the goal of shaping the conscious subject within the three-dimensional organic realm. Once this process is finalized and subjects become fully aware of themselves, or enlightened, and then light up with their enlightenment the unconditionality of divine consciousness, which enables the universe itself into existence. Therefore, only in enlightenment does existence become real, like a personal event. Up until that point, the universe is impersonal. It actualizes itself in the sense of existence only through an enlightened individual - never before and in no other way.

However, the average experience of subjectivity known to us all, average people, does not reflect this ideal state. The majority of subjects experience the early stages of this maturing. We have seen the way in which surrounding space–time on Earth shapes the physical body of man, thus conditioning their movement (destiny). The principle of life (*jiva*) crystalizes through all organic forms, starting with plants then insects, arthropods, mollusks, fish, amphibians, animals that live on land, and, at the end of its evolutionary process, the human form. During this evolutive growth, they were gathering impressions of life and the experience of existence. Their incarnations had an automatic or spontaneous ("immaculate") flow through animal forms, whereas upon entering the human form, the drama of trying out all oppositions and experiences begins to take place. The impersonal process that had been taking place before entering the human form starts to form into a conscious whole, which is called personality. That is why it is followed by the dramatics of a conscious confrontation and responsible participation in the overall phenomena, unlike the previous unconscious submission and spontaneous, natural reactions. All existing culture is an expression of the human aspiration to rise above that level and cross over from the unconscious reactions to conscious actions. Therefore, not every man is completely a man. It is one of the basic reasons why not every man is able to accept and comprehend astrology.

Each of us born into this world grows to perfect the consciousness of our soul, until the moment we become able to manifest divine consciousness in this world while still residing in the body. We are all in some phase of growth along that path. The moment of our birth was determined in accordance with this growth; we are born at the exact moment when our time facilitates our further development. The only reason why our birth takes place is because of the law of karmic maturing in consciousness. The quality of consciousness in one life determines the moment of incarnation of the next one, and with it, the destiny of our new life. The moment of our birth, the astrological natal chart, like a mirror, reflects the true face of our karma. This body we live in and the destiny we go through best reflect the quality of our previous consciousness, the one we entered this body with, following the law of causality, and they are the only real foundation for overcoming the state we are in.

The reality we were born in, which can clearly be seen by reading the natal chart with all its details, is the only true starting point for advancing our awareness. Thinking that anything was wrong in regard to us being born is a negation of this truth and an avoidance of responsibility for putting more effort on raising the level of consciousness and trying to find justification for that in the outer circumstances rather than the inner obstacles. If the conditions of life are appalling, it is only an invitation for us to set them right, the added motivation for working on ourselves to change the immediate surroundings we find ourselves in, and not a proof that something is wrong and we were "unlucky" at birth. Luck has nothing to do with the work on the raising consciousness of our souls. It is more like the tree growing, where every cell upgrades the preceding one in its perpetual growth toward the source of life.

The horoscope has always been a helping hand to any spiritual practice, primarily the one talked about by G.I. Gurdjieff and P.D. Ouspensky, which is the *Fourth Way*.[10] It is a spiritual practice that refuses any form of spiritual techniques and takes existence itself for what it is at any given moment as the only starting point for working on oneself and one's self-knowing. Horoscope offers understanding for the only grounds of the awakening in this life, which is the moment of our birth. It is the moment we stopped on our way to our soul, making it the

---

[10] P. D. Ouspensky: *The Fourth Way*; P. D. Ouspensky: *In Search of the Miraculous, Fragments of an Unknown Teaching*.

only appropriate spot for persevering in our future quest, pointing in the direction of what aspects of our existence we need to work on, what we have achieved so far, and what still remains to be done. Only at the moment when we accept ourselves for who we really are, the way the horoscope shows, will we begin to awaken. If we look for someone or something to shift the blame onto, we are still in a deep sleep while at the same time thinking that we are awake. One aspect of this comatose state of sleep is rejecting astrology altogether. Above all, it gets rejected because an immature man is unable to cope with the chief principles of reincarnation and karma as well as the need to work on themselves. The one who falls in line with these principles without a doubt accepts the laws of astrology as the best assistance one could get on the road to enlightenment, like a map every soul is given not to have to wander about in this world.

Interpreting *karma* as "our" actions and initiating certain processes that will result in corresponding consequences is an expression of intuitive insight that human endeavor is not only a matter of natural causality but something much greater than that. This is a true fact. In man, the activities of a free soul and natural causality intersect. This intersection crystalizes the human personality and becomes whole to the point of overcoming the natural causality and goes on acting in the best interest of the soul that initiates the nature itself into existence. In both cases, whether it is natural causality at work or consciousness of the soul, no ego exists that could be attributed to this activity. The body is just a place where nature and divine consciousness intersect. Nature morphs into its divine outcome, making it seem that this place, our body, acts on its own accord, as some ego. It is an illusion, the same way that metal shavings appear to move toward the magnet of their own free will. Therefore, *karma* is action that does not belong to us but is entirely dependent on the natural processes.

To understand the answer to the question "Why was I born at this specific time with a destiny such as mine is?" We must clarify that "we" were not born, only the body was, which is a natural creation we are identified with as a result of our ignorance. When "our" bodies were born, this did not signify our birth, because we are no different from the same timeless existence or the beingness that always is. By birth, we have simply become identified with the three-dimensional phenomena of the organic world, when we began to observe the physical world with the eyes of this body and experience the environment through its senses. We, then, from the state of unconditioned, divine consciousness of the soul, became contracted and molded into the tiny three-dimensional form this body provides (so contracted and molded that we have forgotten about our soul) but with a clear goal to enable the unconditioned divine consciousness with the concrete, human shape through our soul and body, to stop being impersonal, viewing existence for the sake of existence.

With the oblivion of our soul, we automatically gravitate towards organic phenomena. The soul's traits, expressed in the natal chart (destiny), perfectly match the degree of our oblivion.

This congruence is exactly that which enables the self-knowledge.

By forgetting about our eternal soul, we induce the birth in the body in the space–time of this kind. We let ourselves in for experiencing destiny, which reflects our unconsciousness like in a mirror, and by doing so, it enables the self-knowledge the moment we divert from objectivation and begin to turn toward ourselves. Taking this into account, it appears that every single life leads to self-knowledge. We are born in the body only because we have not done what should be done in order to overcome the natural causality and enable the authenticity of our soul. Every embodiment that happens to us happens in the exact way in which we failed our spiritual outcome and searched for sanctuary in the organic nature instead. Our lives are the accurate manifestation of our level of identification with the being and unconsciousness of the soul that enables it all. Finally, when after sufficient experience and adversity we become aware that our true origination is not in the natural renewals and that we would do better than to look for the salvation in them - but the true foundation of nature is in the soul, which enables it and which is the stuff we are made of - at that point, on that spot, we have transcended our destiny, and our lives are authentic and free; it turns into something universal and eternal and not individual any more, which is only limited and transient.

Therefore, our soul is our essence, enabling the very existence of nature; it is unconditioned by space and time and free from any psychophysical shaping of nature. Owing to this essence of ours, we are able to be conscious and psychically objective because the soul that enables the consciousness goes beyond the psyche and the body altogether. In identification, objectivity is not possible; it can happen only when the principle of overcoming is applied. However, the primordial unconditionality and soul's independence (enabling all of nature) is an impersonal fact, hence its effort to realize itself both personally and concretely. This very aspiration shapes human beings; it is a process that is realized at a different pace in different individuals, in accordance with space and time (i.e., the nature of the three-dimensional reality it is realized in). This process is what we call destiny. It manifests the measure to which the consciousness of the soul has become concretely present and conscious within an individual. This is a process of conformation of nature and not the soul. The soul merely has the power to attract; the soul is not pulled in by the gravity of natural phenomena. Nature conforms according to the divine consciousness of the soul and not divine consciousness in regard to nature. Human nature (and destiny too) is the point of crossing between the soul and the nature, the meaning and the existence, the form and the awareness of it, the freedom and the necessity - their interaction and embodiment. The human, in essence, is nothing at all; only a measure of conforming personification of the presence of soul in a being, the place where the sense of divine and nature are measured, where meaning and existence receive personification for the first time as a whole personality. **Human personality becomes whole only to the point man becomes aware of the meaning and relationship between divine consciousness and nature and to the point they are able to connect the divine consciousness and nature within themselves.**

This same process can be understood if we describe it by saying that our immortal and unborn soul becomes glued to the sticky substance of the natural phenomena of the body and mind and seduced by their gravity to the same degree that we have identified our essence with the natural phenomena. Astrology shows the details of all gravitational pulls in our cosmic surrounding, which set the world in motion, more precisely, the way a soul identifies with the phenomena - the characteristics of the phenomena where our destiny is the direct proof of our oblivion of the spiritual authenticity. Our oblivion of the soul becomes embodied as the psychophysical destiny. Natural determination, like in a mirror, reflects the oblivion of the being. To the degree we are unconscious of ourselves and our true nature, we put up with destiny like a bad dream.

Additional creative thinking will help us to comprehend this further. Nature aspires toward its final goal, which is divine consciousness that enables the existence of nature. In this aspiration, it perfects its ability to act as well as perceive. The ultimate reach of the perfection of this kind is a human being. The essence of the human being is not manifested nature but the divine consciousness of the soul that enables everything; it is the invisible center that has the attracting effect to shape the nature into the visible body and its destiny. The psychophysical beingness (body) is fully wrapped around the presence of the soul (Selfhood); like a magnet, it pulls toward itself the shaping of nature, which has the soul as its end result. *When you apply the visible material of nature on the invisible soul, it consolidates as our overall nature, as the physical form of humans and our destiny.* Selfhood, the human's spiritual core, exceeds space and time. It goes beyond the visible nature but is hidden within the confinement of the body. The body is a suitable place for it to reside in and manifest in nature.

Divine consciousness is omnipresent. It is at the heart of all natural shapes but becomes manifested in the human body in the most personal, that is, the most direct, way. The birth of the body, its movement, and all dramatic experiences it undergoes in the three-dimensional physical world are the movements of nature in its conformation with its spiritual outcome. The body is unconditioned by the dramatics of natural phenomena, although it is humanity's truest nature. The essence of us is always immobile, unconditioned, and exceeding space and time; only the body, which mediates between us and the world, moves along the lines of its destiny because the gravity of natural phenomena drives it accordingly. This is the reason why we can feel the timeless unconditionality of our spiritual being only during the meditative state, hence the bliss and tranquility. All dramatics of natural phenomena (destiny of the body) have been orchestrated in accordance with the quality of the presence of the soul in the being because its presence is not the same everywhere; somewhere it is more mature and expressed, and somewhere it is weaker and beginning to appear slowly. The stronger the presence of the consciousness is, the more favorable the destiny. Consciousness is interaction between soul and nature. By strengthening the presence of consciousness, we strengthen the presence of soul and bring nature one step closer to its sanative goal.

The responsibility for humans to realize their spiritual goal is of utmost importance, given that nature has led the human conformation to its peak by inserting the spark of consciousness within it. Its direct intervention ends at this point. Human consciousness in the body is interaction between soul and nature. This interaction must strengthen itself, and it must design the meaning and purpose of its own existence. This is the process of individuation of personality. Personality is something nature yields as its crop, something that overcomes its causality and something that must integrate all by itself into something unique and new. From that point onward, nature's impact is direct, tempting individuality to develop its independence and consciousness, the same way a mother cannot form her child's personality but can only give birth to it, nurture it, and help it strengthen and integrate its individuality. In the same fashion, nature is unable to do anything for the human once it has provided for their needs; it can only use care and the temptations of destiny to stimulate their growth while encouraging them to develop their personality. It leads them through all temptations to strengthen the presence of consciousness within them, consciousness that is interaction between soul and nature, sense and existence. Work acts as the best tool for its guidance through tempting. Nature, therefore, does not finalize the task; personality achieves the goal. In order to succeed, it must have a relative freedom to act because there is no consciousness without freedom. Human personality has some relative freedom so as to strengthen the interaction between existence and sense, nature and soul, by strengthening the presence of consciousness and becoming the embodiment of the presence of the soul in nature.

Immature individuals use their relative freedom most commonly to do what they like. In doing so, a stroke of fate befalls them because the purpose of the little freedom they have, naturally bestowed upon a person, is to achieve a high level of responsibility for making sense of their own existence.

What does it all practically mean? In reality, this means that an individual is integrated and above the natural causality to the point he or she is aware of the phenomena of life and refrains from reacting unconsciously to various stimuli. An unconscious, spontaneous reaction is the most elementary characteristic of the natural conditionality. The ability to overcome impulsive reactions and understand the true nature of stimuli occurring on the level of the body, feelings, or mind is a characteristic of an integrated personality. Only it has the propensity to act accordingly and creatively, in harmony with the soul that enables everything. All natural reactions we have lived with up until this point were happening on their own accord; our role was passive, sometimes even victim like. In the spiritual sense, we did not even exist prior to this point; there was only nature, doing its thing, speaking and thinking through "our" bodies. Astrology shows us all the details it operated in accordance with, and this knowledge helps us achieve differentiation and actualize the primeval independence of the soul in us.

Much like using the biofeedback device, learning of the astrological influences reveals our weak karmic spots and stimulates the ones we should become aware of.

If, for instance, we experience sudden outbursts of anger or aggressiveness, astrology will indicate a strong or poorly aspected Mars in our natal chart. Knowing this, we will no longer manifest aggressiveness as before, because we now know it is due to the position of Mars rather than our own disposition. The same goes for all planets constituting our character and temperament for all houses that reflect our life choices and orientation. This is one of the ways in which we develop the differentiation of our personality away from the natural urges. Every time we resist implicating ourselves in the elementals of nature testing us, we have become the masters of our lower drives and no longer their slaves, we have strengthened the independence of our personalities, and we have begun to take responsibility for it, taking one step closer to our unconditioned soul. The natal chart reflects, down to the last detail, the way in which nature acts through us in a unique way through body and mind. This provides us with a novel tool of being able to recognize the natural conditioning and taking an objective stance toward it – which means not following it blindly. In this manner, we develop the presence of consciousness and responsibility for the meaning of existence. To be able to do that, we must first be well equipped with the knowledge astrology gives us, and we must know the details of our horoscope, the challenging and prominent factors of the chart, because they are the ones that attract us the most and cause us to be identified with them; in other words, they condition us the most. The sign we were born under is the first and the biggest challenge we must overcome, then the house the sun is placed within, followed by the challenging aspects as well as the remainder of the chart. (Favorable aspects help us become aware and overcome the unfavorable factors.)

Unlike many other spiritual traditions and practices that delve into the depths of transcendence of natural conditioning, astrology has the added bonus of providing us with the insight into our individual and unique situation - objectively and without any ideological, mythological, and collective convictions, it sketches the pattern that illustrates all points of our identification with the natural phenomena that trap us, making this scheme our escape route out of the maze at the same time.

We can divide people according to their level of karmic maturity, the level of consciousness, which is the level they are identified with because of natural causality, which forces them to react unconsciously to events around them, and by doing so, they generate future events. Most people are completely enslaved by impressions and incentives that always induce spontaneous reactions in them; they are unable to differentiate their consciousness away from this chain of causality, and they are even unaware of the ability to do so. Their destiny is completely predetermined, like an apple falling off a tree to the ground. Some people are able to overcome some forms of conditionality and have an active influence over their life to a certain degree but not in its entirety. They can alter their destiny to a smaller degree occasionally but never completely. There are too few people who are able to attain the complete independence of soul from the natural conditioning that takes places during their life in the body. They are awakened and have exceeded the time in which destinies take place. They are free because they see that destiny did not belong to them; it was fully in the hands of nature.

Becoming whole through interaction between soul and nature has two phases. The first one is self-knowing, the cognition of one's own spiritual goal and the independence of the human soul. It is achieved by research and becoming aware of the distinction between one's spiritual essence and the natural influences of any kind of destiny that natural motion generates. Once it becomes properly accomplished, it is evident that the overall nature aspires toward divine consciousness as its result, it attains its salvation through an enlightened man who has reached the ultimate state of self-knowledge. Nature exists for our self-knowledge and freedom. It is the second phase where the utter acceptance of all natural shapes takes place, where we support all of life in deep, self-giving love. Although they appear contrary to one another, both phases are necessary for accomplishing wholeness. Once the first phase has been adequately implemented, the second phase automatically follows. If self-knowledge is not accompanied by loving acceptance of the overall existence, with all its oppositions, and divine consciousness that enables the potentiality of existence and is something that all the living beings aspire toward, it means that something completely different has taken place. Our spiritual essence is the goal that all of life and nature strives for; consequently, our spiritual self-knowledge we confirm by accepting all of life and all of existence. *Love expressed toward everything living and existing is the only way to manifest divine consciousness that enables everything into existence, constituting our essence, or soul.*

The life and destiny of the physical body are the only ways in which we can become aware of ourselves, our Self, which overcomes the body and its birth and death. This path must always be individual because the awareness of the Self can be actualized only in an integrated, whole personality, and personality can be this only if it is unique. This is the reason why it can be shaped only in the three-dimensional world which is the only place where this can happen. Uniqueness, as a prerequisite for the awareness of oneself and the integration of personality, is enabled by constant motion of the celestial bodies, which is never the same. The body has destiny only because it moves through the three-dimensional physical world, which created the body in linear time, and together with it, there is also the mind, which is a reflection of the interaction with other bodies. The mind is a social phenomenon. We tend to experience the deeds of our bodies and minds as our own only to the point where we reach awareness of our unconditioned Self, which is independent of all phenomena and natural necessities. A sufficient level of consciousness will always manifest, at least intuitively, that this body is not our body, as we are inclined to believe, because we are unable to control a larger part of its processes and not even aware of them. A more alert insight will indicate that everything the body does, or is done to or with it, is not ours entirely. In the hidden depths of the soul, each person must admit to himself or herself that they feel trapped in their body like in some biological machine that oppresses them and does more of what it wants than what the intentions of the individual are. Everyone has committed an evil act during their lifetime, and each of us know that we were not evil of our own accord. "Something" drove us to do evil, if nothing else then a misfortunate set of circumstances.

Nobody is guilty for what they have done, because these are all the machinations of nature. Humans can only be held responsible for the good they failed to do and was in the position to do or for not having done the only thing they are supposed to do, and that is raising the level of awareness of their being and openness for the soul that enables the being. All of human suffering is based on a guilt of this kind.

When astrology facilitates our direct insight into things as they are, without the assistance of morality and various convictions, we are one step closer to the insight that our own fiber is different from psychophysical phenomena. This insight can be accelerated considerably by some of the ancient or modern practices of awareness (meditation) that have reached us through many ages of religious and mystical traditions. This helps us attain an understanding of the background of all actions. If such an understanding is lost, the world is viewed from an upside-down perspective; consequently, our actions are opposite of what they should be.

The fact that the soul is different from the natural phenomena is not the same as two entities being different. The soul is the goal of the overall phenomena, and their differences can only be seen in the form of a process and its accomplished goal. Here, the issue is that of time. Time that separates the soul from the natural phenomena can be reduced to the work one puts in to attain awareness. Time is relative; it can be stretched into spaciousness and the shaping of events, and from that perspective, the surfaces of planets can be seen as linear time and destiny, which unravels "over the course of time." It can also be contracted into a personal event and consciousness that is here and now. That is why the quality of the presence of consciousness determines the nature of time, together with it spatial shaping, otherwise known as destiny. Space and time of all phenomena get contracted into consciousness and spread out as the objectively shaped cosmos. This means that destiny depends on the presence of consciousness and not some specific actions; the very presence of consciousness determines the speed of events taking place and their nature. Apart from that, the moment somebody will be born is in tune with the quality of the presence of their consciousness.

This further means that there is no objective redirecting of destiny at will but that the only solution for each individual destiny is to become aware of the subject that is in possession of the specific destiny as well as their position as the reference point in the events that take place in their life and for the sake of them - the one who is being aware.

It has been empirically proven that the best solution for every destiny is the complete overcoming of it and achieving its transcendence in the divine consciousness of our soul. Any different intervention on the destiny, its alterations for the sake of better results, has, so far, yielded poor results entirely. It is the same as trying to establish as good a life in prison as possible instead of breaking free from it.

Any attempt to change one's destiny in some segments of life in order to fulfill one's desires is simply a trap for the new form of slavery and new shaping of destiny. Destiny should be modified from the same level it occurs, from the viewpoint of the body and mind. One should firstly transcend the body and mind, and once we have reached the level of understanding each individual destiny perfectly, we will be able to access the tools for upgrading. However, then we will not create a new destiny, but we will liberate ourselves from any shackles of destiny. At that point, we will be able to create a game of life in accordance with the consciousness of our souls. After this lengthy and demanding process, every life will be a dance of the divine. In reality, with the consciousness of one's transcendental soul, we are able to perceive that every individual life is the divine dance and everyone's destiny is an expression of the divine impulse.

Every destiny is finalized in personality, and each one, in its own unique way, leads to the whole and complete person as its final goal. The purpose of every destiny is, therefore, the shaping of an integrated personality, and every single destiny can be understood only from a perspective of this kind. Personality is personification of the unity of phenomena through which divine consciousness, which enables the overall phenomena into existence, can manifest its personalization.

FREEDOM RESIDES IN THE SOUL

When the experience of existence has gone on long and hard enough through the process of maturing, encountering all opposition throughout its karmic evolution in human form, it finally reaches a moment of facing the meaning of its existence and existence in general. This confrontation is the initial stage of attaining wholeness of personality and is always followed by a crisis. This existential crisis may last for the larger part of life and is often transferred on to multiple lives as well. Up until the point humans reach the state of wholeness as people, they are split into many fragments and oppositions. How and in what way is best determined at the time of their birth, which reflects the state of nature at the moment of their birth because nature and its dynamics are chief creators of their divisions, not themselves. They indirectly facilitate them with their passivity and unwillingness to consciously accept taking part in events, to differentiate themselves from their conditioning.

All phenomena happens to obtain embodiment of their sense in a conscious person, whereas man, whose ultimate goal is not embodiment but to become aware in response to existence, automatically undergoes an existential crisis and does not live a real life. Their life becomes suffering. Suffering is nothing but the pressure imposed by nature forcing consciousness to crystallize, purify, and integrate into a personality, which is the only place where the purpose of existence can be achieved. The final outcome of the whole being is created in the human soul: divine consciousness that enables all, the being too. All of nature aspires toward this goal. ***Only through the human personality is nature completely in harmony with divine consciousness and fully authentic.*** The purpose of natural phenomena is dependent on the personality that has attained the state of wholeness, and that is why everything flows into it, not the other way around; personality ceases to be personality when it is a slave to natural phenomena. Humans suffer when they upset the primeval aspiration of all life and if they seek their stronghold in nature through the objectivation of its phenomena, in objects, in the slavery to linear time where the transience of everything is experienced. They then go against the natural flow, and it is no wonder that nature throws low blows at them and disrupts their destiny because that is the way it diverts them from the wrong path and away from the identification with objects, driving them instead toward finding the only stronghold in themselves, in integrating their mind with the consciousness of the soul and in becoming the embodiment of divine consciousness, which gives life and authenticity to everything living while saving all beings in the process. The singular destiny of the human is to be the savior of being, to enable life with consciousness and love and not to end up as its slave. Awakened humans realize that they are always in the consciousness of their soul and there is no need to make themselves authentic (to "save" themselves), but the overall nature heals itself and becomes authentic through them when they are awakened.

The meaning of existence is in its awareness, and awareness is nothing but the response to this primeval aspiration of the existence itself. If there is no response, existence is in vain; it appears destructive and painful. The essentiality of consciousness is in the dialogue or communication with the existence. Without the personal participation in existence, there could be neither awareness nor the meaning of existence. We participate personally only when we do not act and react spontaneously in accordance with the impulses coming from Mars, the moon, and other factors of natural conditioning that astrology shows in detail; instead, we manage to resist them and treat them objectively, in the best interest of the soul and not the urge. Every time we pull it off, the path to freedom is wide open. Only triumphs of this kind can liberate us. Without the personal participation, destiny is like a car in motion with the driver fast asleep. It is hardly a surprise when a destiny of this kind ends tragically. Smaller bumps in the road during such an unconscious ride should wake up the driver while there still is time in their physical life. If they crash against the end of their life path while still being asleep, they will go on snoozing through their next life. The one who sleeps all their life long and always complains about their ill fate should do better to wake up and see where they are headed.

For as long as personality is divided, its dialogue with the existence is divided, and the answers to the issues of life come across are rarely satisfactory. Only an integrated person has an integrated relationship with life and the meaning of it. The first experiences of the integrated relationship with life we acquire in our relationships with the people whom we are close with or with the person we love. However, it does not have to be with loved ones only but all people we have a meaningful relationship with. We learn to love every person we understand deeply enough over time. There is no discrimination here; this person could be our enemy. Love is the only confirmation of the proper understanding of other people but also the world in general. With the consciousness of our minds, we can see a person the way he or she is but always objectively, from the outside, while we can only experience them with love for what they truly are inside. Love is deeper than the consciousness of the mind. Where the consciousness of the mind reaches its limit, love moves on. The reason for this is that love comes directly from the higher consciousness of the soul and not from the mind, which is of the body.

In the dialogue with other people and through our love toward them, we learn to communicate with wholeness that enables the overall life into existence. The wholeness is in all of us, and by knowing other people, we eventually know it. Then this "other" disappears, and only the self-knowledge remains. The one can neither be known directly in other people nor through other people. The other was merely a reflection of the one, a motivation to converse with it. This one is the core of our being, and only can we consciously actualize it, as our essentiality, uniquely and remarkably.

Love is the feeling of unison with the being, its complete acceptance. This unison we must first find within ourselves. We must learn to develop the power of unbiased perception of all feelings in the body in a systematic fashion from the very point of their origin, both pleasant and unpleasant ones. Once a feeling is fully developed and consumes us completely, it is too late to observe it then. When we learn to detect it on the finest level, at the point of its origin, we will not become a slave to it; we will instead see the causes for this feeling in our environment, in ourselves, or on some planet transiting our horoscope chart. When we learn this first step, the impartial perception of impulses within the body, feelings, or mind, the way they are in the beginning, we are ready to make the second step, which is to avoid reacting spontaneously to them. If we were to react spontaneously to them, we would be conditioned by them; we would not be any different from the impulses and their contents.[11]

The experience clearly shows that existence can never be experienced simply by trying to envisage it, because we would then see what we design with our minds. Existence can only be experienced directly, without any devising. When nothing is interpreted, no wrong can be seen, only things the way they really are. In reality, there is nothing hidden away from us. The human mind is the only entity in the whole of nature that does not see things for what they are; he must painstakingly discover them. Humans become enlightened when they realize there are no objective obstacles of any kind for their mind to comprehend, but the mind itself is an obstacle for their being to recognize the absolute as their innermost substance. There are no differences or borders between the individual and the absolute being; the only difference is the one orchestrated by the mind in its imaginings. The mind is always split and narrowed down to something singular, while the being is always whole. The human being already knows everything because it is composed of everything; it is the embodiment of all principles and dimen-

---

[11] On the most general principles and details of the way the being can be experienced and made aware of, see the final chapter of my book: "Sâmkhya – An ancient science of nature and human soul'. Afterward, William Hart: *The Art of Living Vipassana Meditation*, according to the teaching of S.N.Goenka. Further reading should be: Nyanaponika Thera: *Buddhist Meditation*. On meditation see my book: "*Meditation. The First and the Last Step. From Understanding to Practice*".

sions of nature. To experience existence directly for what it is means to experience it with one's whole being and complete acceptance at any given moment - with love. There is no other way.

Destiny is a natural process, and it keeps conditioning us until we have sufficient maturity to consciously and directly face the phenomena for what they are and learn to distinguish our dissimilarity from them. Then it stops. The horoscope interpretation is a very illustrative example. When, with the help of astrology, we become aware of some situation, whether it has to do with our character or something that is going to happen, its realization changes. The fact that we have become conscious of it has brought change in the process of realization. If we had not become aware of it, if we had not analyzed the horoscope, it would have inevitably happened to us as our destiny. The very presence of consciousness has changed the nature of events and turned them into a challenge for raising awareness. If we accept the challenge, it is no longer an inevitable destiny we have to deal with but a part of our maturity that has increased the level of our consciousness. By doing so, it changed our position in the overall phenomena. The same way there are chemical processes that can only take place in full sunlight, never in the dark, the very presence of the light of consciousness changes the processes of both inner and outer events; it speeds up the time of their accomplishment and perfects their expression.

Only when astrology is used in this way can it show its true value and supreme quality. It becomes an irreplaceable tool of self-awareness. Its primary function was never divination but for the self-knowledge of the one who has been allotted a certain destiny to facilitate their liberation from it.

The purpose of culture and spirituality is for humans **to stop searching for their authenticity outside of themselves** via outer emissaries, religion, or any similar systems of beliefs, rituals, myths, symbols and authorities but instead, in some objectivization, **to find it inside, within the realms of their own being**, directly and in their own maturity, during the personal confrontation with the existence, which is always our Selfhood and authenticity, and to learn to take part responsibly in everything it entails. (Luke 17:20–21)

Destiny exists only for those who are not whole, who do not accept with love everything that exists, who refuse to accept even the tiniest life form, or who force themselves to accept it by means of some conviction. Such a person objectivizes the meaning of existence and looks for solutions in objective changes. That is why objective nature has such a conditioning effect on them, because they are identified with it. Only a whole person is aware of its Self, its soul that is one with the divine consciousness that enables everything into existence. For those who wish to "know their destiny," by wanting to do so, they confirm they that do not know who they are or what life is. The one who is self-aware realizes that all of nature is one, whatever is going on in or with it, and that the one in question is their being that is independent from singular events. He becomes aware of the fact that the whole of nature is enabled by the divine consciousness, making it automatically an expression of divine consciousness, and there is no conditionality or destiny in it. Such a concoction may only exist in the mind that is deeply unaware of this authentic state of being because he only sees certain aspects of existence. Indeed, some aspects of nature reflect discrepancy and disharmony. When man is identified with them, he experiences suffering and destiny; however, when existence is viewed as one wholeness, it can only be interpreted as pure goodness and a reflection of divine consciousness. For humans to be able to see existence in the way described, as pure goodness in all its aspects, they must be whole themselves. The state of their existence will correlate with existence in general.

Such a human is no longer under the influence of their horoscope chart; nature does not enslave or condition their evolution anymore. It no longer has any unfinished business with them, the task has been completed, they are now independent, and they have become a person, a human. Up until that point, they were something in need of shaping, something Mother Nature needs to nurture and take care of. She has been doing so with all oppositions, both mercilessly and softly, nicely and discomfortingly, killing and bringing humans back to life many times over till they realized that they are not just a body and that they are answerable for their consciousness, which they must never lose on account of events coming from the outside world. They can keep it only if they find the only stronghold within themselves. Once he achieves this, the situation changes miraculously: Since all of nature is on a constant quest to be healed and saved through an enlightened man, when he consciously responds to this aspiration, turning toward their own integration, the whole of nature begins to assist and even becomes obedient to the point that for an onlooker, it seems that man is suddenly in possession of some supernatural powers, which he uses to give and promote life while asking for nothing in return. At that point, man begins to experience existence as consciousness and bliss. All individual aspects of their existence, objectively displayed by astrological principles of their natal chart, both good and bad, he begins to experience as diverse stimuli on the path to awareness and not as something "good" or "evil." He sees any potential of destiny, or some current situation, as nothing but an invitation to raise the level of their awareness. Consequently, they are not inclined to start changing anything in the outside world, because he knows that situations in the outside world change automatically when the state of awareness of the subject who experiences them shifts. He sees that everything is following a set path and will eventually reach the finish line because everything belongs to the whole; nothing is ever outside of the range of divine consciousness, and therefore he does not worry about anything. He realizes that the purpose of all events is to generate inner maturity and consciousness. There is no point in trying to change the outside world beforehand. The quality of outer circumstances can be changed only with personal maturity. There is no other way. Now man has become awakened, whole, and the embodiment of divine consciousness and unconditionality that enables everything. The wholeness is independent from events within its domain. In the same way, an enlightened human, as the personification of the whole, is independent from all events within nature itself. The psychodynamics of space–time no longer affect them, and their gravity, which creates destiny, no longer applies to the phenomena of their life, because the gravity of their Selfhood is the one that has overpowered the environment. In that respect, they resemble the sun. They ceases to seek stronghold in the creature that has been created, because they have found it in the divine, which enables the being into existence. They are no longer the object of their own destiny but the one who creates and radiates life about them because the divine consciousness that enables everything now freely emanates from them.

They are then free from destiny but not in the sense that they are its "master" and are at liberty to do as they see fit. Quite the contrary. They do not have a goal to strive for as there are no objects in their life, because they are the goal of everything living. They are the final result of everything and accept all beings with love as they are. They heal by their very existence; namely by their mere presence in this reality and the kindness they show to their fellow human.

Printed in Great Britain
by Amazon